# SOILS IN CANADA

EDITED BY R. F. LEGGET

This work originated in a Symposium forming part of the programme for Section IV (Geological Sciences including Mineralogy) of the Royal Society of Canada, which met at Queen's University, Kingston, in 1960. Of wide scope, it demonstrates the progress now being made in Canada in the study of its vast area of soils. The papers of this Symposium are unique in that they present for the first time a combined picture of three aspects of soil science—the geological, the pedological (or agricultural), and the engineering (known as Soil Mechanics). The book serves, of course, mainly as an introduction to a large subject, but some more detailed papers give an idea of the depth as well as the wide range of soil studies in Canada today.

The contents can be summarized as follows. First, come seven papers on Pleistocene geology in Canada, followed by a study of muskeg (which forms half a million square miles of Canada's surface) and one of soil mineralogy. Four papers— one general and three regional—of pedological interest follow. Finally, come four papers on soil mechanics: one relating agricultural and engineering soil studies; one discussing geology's influence on the siting and building of airports; a detailed account of the properties of Leda clay; and a general review of the soil problems facing the Canadian civil engineer.

R. F. LEGGET, F.R.S.C., is Director of the Division of Building Research at the National Research Council, Ottawa.

# THE ROYAL SOCIETY OF CANADA

## Special Publications

# SOILS
# IN CANADA

Geological, Pedological, and
Engineering Studies

THE ROYAL SOCIETY OF CANADA
SPECIAL PUBLICATIONS, NO. 3

Edited by Robert F. Legget

PUBLISHED BY THE UNIVERSITY OF TORONTO PRESS
IN CO-OPERATION WITH
THE ROYAL SOCIETY OF CANADA

# PREFACE

THE PAPERS IN THIS VOLUME were presented as a Symposium which formed part of the programme for Section IV (Geological and Allied Sciences) of the Royal Society of Canada at its annual meeting held at Queen's University, Kingston, Ontario, in June 1960. Twelve of the papers were presented in summary form at two sessions on June 7th and 8th, the other papers being presented by title only. Professor F. F. Osborne, the President of Section IV, presided as Chairman of these sessions.

The Programme Committee responsible for the Kingston meeting was headed by Dr. J. E. Hawley of Queen's University, the members being Dr. A. W. Jolliffe, Dr. Y. Fortier, Dr. J. W. Ambrose, and Mr. R. F. Legget, assisted by Dr. S. C. Robinson, Secretary of Section IV. The planning of the Symposium was worked out, in co-operation with the Programme Committee, by a small group consisting of Mr. C. B. Crawford, Dr. N. R. Gadd, Dr. A. Leahey, Dr. V. K. Prest, and Dr. P. O. Ripley, with Mr. R. F. Legget acting as convener.

Appreciation is recorded for the assistance given by all those mentioned and by the authors in the development of this Symposium.

The Editor is grateful for the assistance he has received from Mr. C. B. Crawford and Miss J. Butler in seeing the volume through the press and also for the continued interest of Miss F. G. Halpenny, Editor, and the expert assistance of Mr. R. I. K. Davidson, Assistant Editor, University of Toronto Press.

NOTE TO SECOND PRINTING

For this second printing, an Index has been added in accordance with suggestions in some of the reviews of the original edition. The Index has been prepared with the assistance of all the authors. Two brief addenda have been made on pages 21 and 96, and some minor errors have been corrected. The Editor would welcome suggestions for further improvements in the event that another reprinting of the volume is made necessary.

R.F.L.

# CONTENTS

# CONTRIBUTORS

J. E. ARMSTRONG, Senior Geologist, Geological Survey of Canada, Department of Mines & Technical Surveys, Vancouver, B.C.

L. A. BAYROCK, Earth Sciences Branch, Research Council of Alberta, Edmonton, Alberta.

W. E. BOWSER, Senior Pedologist, Canada Department of Agriculture, University of Alberta, Edmonton, Alberta.

J. E. BRYDON, Research Officer, Mineralogy Section, Soil Research Institute, Canada Department of Agriculture, Ottawa, Ontario.

H. L. CAMERON, Director, Photogrammetry and Geology, Nova Scotia Research Foundation, Halifax, Nova Scotia.

C. B. CRAWFORD, Head, Soil Mechanics Section, Division of Building Research, National Research Council, Ottawa, Ontario.

A. DREIMANIS, Department of Geology, University of Western Ontario, London, Ontario.

J. A. ELSON, Department of Geological Sciences, McGill University, Montreal, P.Q.

S. A. FORMAN, Head, Soils Mineralogy Section, Soil Research Institute, Canada Department of Agriculture, Ottawa, Ontario.

C. P. GRAVENOR, Peace River Mining and Smelting Ltd., Edmonton, Alberta.

R. M. HARDY, Dean, Faculty of Engineering, University of Alberta, Edmonton, Alberta.

P. F. KARROW, Department of Civil Engineering, University of Waterloo, Waterloo, Ontario.

A. LEAHEY, Associate Director of Program (Soils), Canada Department of Agriculture, Ottawa, Ontario.

R. F. LEGGET, Director, Division of Building Research, National Research Council, Ottawa, Ontario.

N. W. MCLEOD, Engineering Consultant, Department of Transport, Ottawa, Ontario.

V. K. PREST, Pleistocene Section, Geological Survey of Canada, Department of Mines & Technical Surveys, Ottawa, Ontario.

N. W. RADFORTH, Chairman, Department of Biology, McMaster University, Hamilton, Ontario.

N. R. RICHARDS, Head, Department of Soil Science, Ontario Agricultural College, Guelph, Ontario.

A. RUTKA, Materials & Research Section, Ontario Department of Highways, Downsview, Ontario.

P. C. STOBBE, Director, Soil Research Institute, Canada Department of Agriculture, Ottawa, Ontario.

# SOILS IN CANADA

# INTRODUCTION

## R. F. Legget, F.R.S.C.

THE TERRAIN OF CANADA, the third largest country of the world, is unique in that it contains the largest individual area of glacial soils of any country and correspondingly the largest area of enclosed fresh water. In addition, it is distinguished by at least 500,000 square miles of muskeg. When these facts are related to the relatively small population of the country (about five per square mile) it is perhaps not surprising that so little progress has been made in the study of the varied aspects of the land that is Canada. In recent years, however, this picture has been changing and real progress is now being made in geological, pedological and engineering soil studies in Canada.

Some of the earliest Canadian geologists, including Sir William Logan and Sir William Dawson, made notable studies of Pleistocene phenomena; there is therefore a long tradition of soil study in geological circles. Soil mapping for agricultural purposes was started about forty years ago and has made steady progress, greatly accelerated in recent years. Attention to the engineering properties of soils has been a feature of individual Canadian engineering projects for many decades. Some Canadians have been pioneers in this field, notably Samuel Fortier, a graduate of McGill, who did much notable engineering work in the Western States at the end of the last century. Instruction in the new discipline of Soil Mechanics was first given in a Canadian university as early as 1932, several years before it had been officially recognized as an independent branch of study.

In the years since the end of the Second World War, the Geological Survey of Canada has established a Pleistocene Section and has carried out Pleistocene field investigations in all parts of the country. Soil surveys for agricultural purposes have progressed in all provinces by the co-operative efforts of the provincial and the federal governments and the colleges of agriculture. Soil mechanics has achieved further recognition not only in educational services but in application on almost all major civil engineering work now carried out in Canada.

With these concurrent developments the liaison between workers in the three disciplines has progressed fruitfully. As an example of this trend, it may be noted that the Associate Committee on Soil and Snow Mechanics of the National Research Council (the function of which is to stimulate and co-ordinate research into the terrain of Canada, with special reference to engineering problems) has for long included in its membership an agricultural soil scientist and at least one geologist. The links between the three

3

approaches to soil studies were demonstrated publicly by papers and discussions at the Ninth Annual Canadian Soil Mechanics Conference held in Vancouver in 1955. The liaison is further demonstrated by the appearance of this volume containing papers presented to the Annual Meeting of the Royal Society of Canada, Section IV, held at Kingston in June 1960.

This liaison between geological, pedological and engineering soil studies in Canada has developed despite a rather important difference in terminological practice. The agricultural soil scientists have their own definition for the word "soil"; as is well known this includes what others call "topsoil" and the well-recognized "B" and "C" horizons. Geologists and engineers, on the other hand, following accepted practice established well over a century ago, use the word "soil" to indicate all the fragmentary material in the earth's crust overlying bedrock, with the possible exclusion of the thin layer of topsoil (generally so described) which is of such importance in agriculture.

There has been in Canada mutual acceptance of this variation in the use of the word "soil" and corresponding mutual respect for its varying use by the three disciplines concerned. It may be regretted that two words have not come into general use in the English language to correspond with the two Russian words used to distinguish the soil that is of interest in agriculture from the more general soil considered by the geologist and the engineer. In the absence of this convenience in terminology, the word "soil" is therefore used in this volume in the two distinct ways but its precise meaning will always be evident from the context in which it is used.

There are other differences in terminology between the practices of the three disciplines represented. As an example, it may be noted that geologists use the word "compaction" where workers in the field of soil mechanics would use the word "consolidation," each for the same phenomenon. As these terms are used in this volume, however, the context always makes clear the meaning intended. It is to be hoped that with time and with steadily increasing liaison between geologists and engineers working together on soil problems, these differences will tend to disappear. Correspondingly, it is to be hoped that eventual uniformity will be reached in the limiting equivalent grain sizes for the subdivision of soil particles into sand, silt, and clay sizes, in which slight differences of practice still exist, differences that may be noted by the careful reader of this volume even though, again, the context obviates any possibility of confusion.

Canadian workers in all three fields of soil study are appreciative of the overlapping of their respective interests and therefore of the value to be derived from mutual discussion of soil problems. This is reflected by the contents of this volume which deals with the soils of Canada from the three points of view. The volume may be regarded merely as an introduction to a vast subject but it is hoped that, through the arrangement of the papers it contains and with the aid of the associated bibliographies, it will present at least a general picture of the major soil types and soil problems encountered in the Dominion. Its coverage is not in any sense complete but it may possibly pave the way for more comprehensive treatments in future years.

The volume starts with a general review of the present state of knowledge regarding the Pleistocene geology of Canada. Against this background six regional geological studies follow, each dealing with a distinct area in which soils with varying geological histories are encountered. The importance of the relatively new study of muskeg, and its significance to Canada, is indicated by the next paper which serves as an introduction to the work already done in this field by its author. All soil studies are dependent to some degree upon soil mineralogy which is therefore the subject of the next paper, a summary being presented of the results of recent research.

Soils are next considered from the pedological viewpoint starting with a broad review of the position of pedological studies in Canada today. Two papers follow dealing with major agricultural soil groups of importance in Canada, followed by a paper describing in general terms the soils met with in an important agricultural region of Ontario.

Four papers dealing with soils from the engineering point of view complete the volume, the first being closely linked with the preceding pedological papers in that it illustrates how the results of agricultural soil studies are being put to use by engineers. A somewhat unusual aspect of the geology of soils is treated in the paper describing the influence of geology on the design and construction of airports. As an example of the detailed studies now being made (in soil mechanics laboratories) of some of the soils that are peculiar to Canada, a summary of research into the properties of the Leda clay is next presented, this being a companion paper to the geological review of the soils of the Champlain Sea. The volume is brought to a close by a general review of soil problems as encountered in civil engineering in Canada and their correlation with geology.

Many of the papers point the way to needed research and several of them demonstrate clearly the interrelation of the different approaches to the study of soil represented within the volume. Presentation of most of the papers at the meeting of the Royal Society provided an interesting example of the cross-fertilization of ideas that can arise from such joint consideration of a great subject. It is therefore hoped that this volume will not only provide a useful reference for those concerned with the soils of Canada, and act also as a guide to those outside Canada who wish to know generally about Canadian soils, but that it will also stimulate further research in all three branches of soil study, the results of which, when co-ordinated, will still further advance general knowledge of the soil that makes up the land that is Canada.

# GEOLOGY OF THE SOILS OF CANADA*

## V. K. Prest, F.R.S.C.

DEFINITION OF THE WORD "SOIL" varies considerably according to the point of view of the scientific discipline concerned. Nevertheless, from any point of view the term "soil" applies to some part or aspect of the unconsolidated material that mantles the bedrocks of Canada. Thus a study of the origin and history of the primary materials comprising this mantle of debris should be of interest both from the standpoint of broad areal distribution and of microscopic mineralogical composition. Geologists are directly concerned with the origin and history of the surface deposits and with many other aspects of this study that have a direct bearing on their main task. Geologists endeavour, therefore, to keep in contact with the progress of work in the fields of pedology and soil mechanics. The present symposium brings together the three major disciplines concerned, that each may present a brief summation of work in its specific field and locale. The geological work is presented first as it is concerned with the basic study of the parent materials. It is appropriate that the broad concepts of the origin of the complex unconsolidated deposits of Canada, their stratigraphy, and their areal distribution should be considered before more detailed discussions.

As is widely known, Canada contains the greatest area of glacial deposits of any country in the world. These deposits have played a profound role in the development of our farm and forest economies. They have been equally important, though less obvious, in the shaping of our mining, fishing, and transportation industries, since the very nature of the terrain, the location and courses of our river systems, and the position and character of seashores are the direct result of complex glacial events. The need for comprehensive studies of these deposits and their origin is therefore great. Though the glacial deposits themselves, and the events that led to their deposition, have been studied haphazardly for over a hundred years, and excellent basic work was done before the turn of the century, it is only in the past ten to twenty years that they have begun to receive the over-all attention they warrant.

The geological study of the surface deposits may be referred to as surficial geology or Pleistocene geology. The latter name, applied by Sir Charles Lyell in 1839 on palaeontological grounds, signifies the "most recent" epoch of the earth's history. At that time, however, the glacial aspects of the Pleistocene epoch were not generally recognized, although it was ten

*Published by permission of the Director, Geological Survey of Canada.

6

years before this that the idea of broad-scale glaciation in Northern Europe had been first announced; two years before this date Louis Agassiz had first formulated the concept of a great ice age for Europe and Northern Asia (Flint, 1957). The Agassiz concept was first known in North America in 1839 but it was not until 1841 that it was given any degree of publicity. General acceptance of the concept or theory of glaciation in North America was assisted by the arrival of Louis Agassiz himself in 1846 to take up a teaching post at Harvard University but opposition to his theory lingered on until the end of the century. These early dates are of special significance on the occasion of this Symposium in view of the fact that Sir William Logan in his report for the year 1845–6 clearly described the effect of glacial action in the Lake Timiskaming Basin of Northern Ontario, and in the vicinity of Ottawa (Logan, 1847). This record of inland glaciation in the far north, as a result of field observations made in 1845 (the year before Agassiz came to America) is assuredly further proof of the outstanding scientific ability of the founder of the Geological Survey of Canada.

## The Pleistocene Epoch

The basic concept of the Pleistocene epoch in both Europe and North America, as one of multiple glaciation involving roughly the last million years, has changed but little over the past few decades. The advent of the $C^{14}$ or radiocarbon method of dating organic materials in the younger Pleistocene deposits, and modern pedological reappraisals of weathering phenomena or soil profile development, give some indication that the time span of the Pleistocene epoch is substantially less than a million years. Nevertheless, four major glaciations separated by three interglacial stages remain inherent in the North American concept of the Pleistocene epoch. Each glacial stage may have involved a period of somewhat less than 100,000 years, from its inception to its end, whereas each interglacial stage may well have been of longer duration. Already the detailed analysis of the last or "Wisconsin" glacial stage in North America, which has received more study than any other stage, has come under attack and several changes or reorganizations have recently been advanced. The radiocarbon dating method has permitted the determination of certain Wisconsin glacial events in North America and has indicated a speed of glacial retreat much greater than formerly visualized. If the advance of a major glacier can take place as fast as the indicated retreat, then the length of the glacial stages at least must be greatly shortened. Possibly, however, there were halts and ice-frontal retreats of some magnitude during the main advance and at the glacial maximum, that represents much longer periods than those necessary for the retreat and disappearance of a continental glacier. A time span at least approaching 100,000 years may therefore still be possible for the Wisconsin glacial stage.

TABLE I
CHRONOLOGY OF THE PLEISTOCENE EPOCH
(from Prest, V.K., 1957)

| Age (stage) (Interglacial ages are set in italic) | Possible duration in years | Sub-age (Substage) | Probable age in years |
|---|---|---|---|
| "Recent" | 7,000 | Period of generally cool, wet climates, marked by glacial advances in some regions | |
| | | (Medithermal) | 0– 4,500 |
| | | Interval of glacier retreat (Altithermal) | 4,500– 7,000 |
| | | Cochrane | 7,000– 8,000 |
| | | Interval of glacier retreat (Timiskaming) | 8,000– 9,500 |
| | | Valders | 9,500– 11,000 |
| | | Interval of glacier retreat (Two Creeks) | 11,000– 12,000 |
| | | Cary | 12,000– 14,500 |
| Wisconsin | 100,000? | Interval of glacier retreat (Brady) | 14,500– 17,000 |
| | | Tazewell | 17,000– 18,000 |
| | | Interval of glacier retreat | 18,000– 20,000 |
| | | Iowan | 20,000– 21,000 |
| | | Interval of glacier retreat | 21,000– 22,000 |
| | | Farmdale | 22,000– 25,000 |
| | | Period of ice accumulation | 25,000– 45,000? |
| | | Probable non-glacial intervals | 45,000– 60,000? |
| | | Period of ice accumulation | 60,000–100,000? |
| *Sangamon* | 125,000? | | |
| Illinoian | 100,000? | | |
| *Yarmouth* | 200,000? | | |
| Kansan | 100,000? | | |
| *Aftonian* | 300,000? | | |
| Nebraskan | 100,000? | | |

## THE EARLY PLEISTOCENE RECORD

The Canadian record of early Pleistocene events and deposits is of general interest. The best evidence of the older glacial stages is clearly the occurrence of indisputable interglacial deposits lying between or beneath glacial deposits. The most famous record in North America is that of the Toronto Interglacial Formation (see Prest, 1957, pp. 448–50). The Don Beds of this formation afford conclusive evidence of a period with a warmer climate than prevails now. A layer of till separates the Don Beds from the underlying bedrock. Here, then, are the deposits of one of the "old" Pleistocene glacial stages and a following interglacial stage. The latter is currently regarded as representing the Sangamon interglacial stage, but it could be older. Other occurrences of buried organic materials in Southern Ontario, although ranging in age back to more than 40,000 or 50,000 years, cannot be related with any certainty to an interglacial stage but may rather represent shorter lived, non-glacial intervals, known as interstadials, of the last or

Wisconsin glacial stage. (These interstadial periods will be discussed later.) The Missinaibi peat beds in Northern Ontario, however, are still an enigma. The climate, as indicated by palynological studies (Terasmae, 1958) was definitely cooler than the present, and the age, according to the late H. deVries of the Groningen radiocarbon laboratory in Holland, is of the order of 55,000 to 65,000 years.[1] Possibly this is the Sangamon period in that area but, in view of the indicated cool climate, as opposed to the concept of warm interglacial periods, even this "old" date is regarded as "advance" Wisconsin. The record left for study however, is fragmentary, there being no evidence of weathering profiles on these buried deposits: a Sangamon age must therefore be admitted as a possibility.

Elsewhere in Canada there is supporting evidence of the older or pre-Wisconsin Pleistocene record in many places. In the Maritimes, although there is as yet no conclusive evidence of interglacial deposits, several occurrences of wood from beneath till over a wide area in the vicinity of the Bras D'Or Lakes, Cape Breton Island, and one example of marine clay and other sediments including peat and wood from between two tills, may well bear evidence of an interglacial stage. The wood from one occurrence was radiocarbon dated at more than 38,000 years (W-157). At the north end of Cape Breton Island near Bay St. Lawrence, W. Neale of the Geological Survey of Canada, has recorded a wave-cut bench about 3 feet above present high-tide level overlain by 30 feet of sand and gravel and a five-foot mantle of till-like material (Prest, 1957, p. 447). Near the base of this section, there is a one-foot layer of silt, gyttja, peat and wood. Pollen studies by Terasmae reveal a plant assemblage that is believed to represent an interglacial period.

In Quebec, not far from Three Rivers, sand and peat comprising the St. Pierre sediments overlie a red till and are in turn overlain by varved sediments and a grey till (Gadd, 1959). Wood from the peat was treated at the Groningen laboratory and H. deVries postulated an age of the order of 60,000 to 65,000 years. The St. Pierre sediments may therefore represent the late Sangamon; but an early or "advance" Wisconsin interstadial must be also considered a possibility.

In Manitoba, there is again no conclusive evidence of interglacial deposits. but fossiliferous sediments from near Duck Mountain west of Lake Winnipegosis (Tyrrell, 1891) may well fall into this category. A fairly recent discovery of an indurated gravel or conglomerate, bearing plant remains, discovered on the Seal River, near the border of the Northwest Territories by F. C. Taylor, may also be interglacial according to Terasmae on the basis of palaeobotanical studies.[2]

In Saskatchewan, southwest of Duck Mountain, Tyrrell (1891) investigated a well from which a new species of larch was obtained at a depth of 200 feet. At this site 8 feet of alluvium overlies 24 feet of drift that Tyrrell

[1] H. deVries, personal communication with O. L. Hughes, 1959.
[2] F. C. Taylor, and J. Terasmae, Geol. Surv., Canada, personal communication.

believed to be a till, and this in turn overlies one foot of sand and 234 feet of soft blue clay with one-inch sand layers every two feet. The Duck Mountain area offers great scope for detailed stratigraphical work. In Saskatchewan, in the Qu'Appelle Valley, stratified deposits also occur beneath till and contain bones and teeth of bison, mammoth, horse, wolf, and bear that may well date from an interglacial period, probably the Sangamon (Christiansen, 1960).

In Alberta there is as yet no proven record of interglacial deposits but buried wood and decomposed organic matter in sand and gravel beneath two tills near Lethbridge, and three tills farther east, may well be interglacial in age according to A. M. Stalker.[3] A radiocarbon dating of wood from the Lethbridge site indicated an age of more than 37,000 years (L. 455A). These organic-bearing sediments have been observed at intervals for over forty miles along the river. Beneath this horizon, two tills have been observed by Stalker in several places, and three tills near Taber, Alberta. The "old" tills are exceptionally dense materials and markedly jointed. In extreme southwest Alberta a succession of Cordilleran tills interfinger with Laurentide tills; it is considered probable that all four glacial stages of the Pleistocene epoch are represented. Biological evidence of an interglacial stage is the occurrence of gastropods and pelecypods from beneath two black soil zones that are separated by a nine-inch bed of volcanic ash that Stalker considers to be "old".[4] One species of pelecypod is known to have been widespread in Yarmouth time.[5] Similarly a species of gastropod found by Stalker in the Red Deer River area in stratified materials beneath two tills is considered indicative of the Yarmouth interglacial stage.

In British Columbia there is no observable section of undoubted interglacial deposits but there are certainly thick deposits of pre-Wisconsin drift sheets and non-glacial deposits. On Vancouver Island, beneath the widespread classical Wisconsin drift sheet, there are non-glacial sediments termed the Quadra from which several samples of a variety of organic materials have been collected; radiocarbon dates obtained range from 25,000 to more than 40,000 years (Fyles, in press). Beneath similar beds near Victoria (dated at more than 40,000 years), there is a till beneath which a succession of gravel, sand and silt together with two or three feet of peat indicates an extensive non-glacial period that is probably interglacial according to Fyles.[6] In the Lower Fraser Valley, the youngest major drift sheet again overlies stratified sediments including organic materials that are considered to be the correlative of the Quadra beds on Vancouver Island. Beneath these again, there is a widespread till of unknown age beneath which drilling has indicated a thick sequence of drift that includes both till and sediments with organic remains. The drift has a total thickness of 2,100 feet (Armstrong,

[3]A. M. Stalker, Geol. Surv., Canada, personal communication.
[4]A. M. Stalker, personal communication.
[5]Shell identifications by Miss F. J. E. Wagner, Geol. Surv., Canada.
[6]J. G. Fyles, personal communication, Geol. Surv., Canada.

1960). In view of their stratigraphic position these drift deposits must include one or more interglacial and early Pleistocene glacial stages.

In the Yukon Territory Pleistocene studies have not progressed to the point where a stratigraphic sequence has been established. There is, however, evidence of an old glaciation and of organic deposits older than at least the last glaciation (Bostock, 1936 and in press). Palynological work on sediments from beneath a lava in Miles Canyon near Whitehorse, reveal a floral assemblage similar to that of the present,[7] and hence probably interglacial. At the junction of the Pelly and Yukon Rivers, one hundred and fifty miles to the north, wood obtained from gravels beneath a lava has been dated by radiocarbon at more than 38,000 years old.

In the Northwest Territories evidence of interglacial deposits has been indicated on both the mainland and Arctic Islands. In the Mackenzie River Valley near Reindeer Depot, interbedded silt and peat are buried beneath more than 200 feet of drift. Palynological studies on samples, collected by E. Porsild of the National Museum of Canada, indicate an assemblage similar to that now growing some fifty to seventy-five miles farther south (Terasmae, 1959). This occurrence is therefore probably an interglacial deposit. B. G. Craig (Craig and Fyles, 1960) indicates a complex history for this part of Northern Canada. He believes that similar deposits underlie the whole Mackenzie Delta and that they may have been overridden by pre-Wisconsin glaciers. Peat from the Mackenzie River site collected by J. Ross Mackay, has been radiocarbon-dated at more than 44,000 years old (L. 522A). Farther south along the Mackenzie River, there are gravels beneath till which Craig is convinced are either interglacial or preglacial deposits.

The first evidence of interglacial deposits on Banks Island came as a result of palynological studies by Terasmae on samples of peaty silts collected in 1952 by T. Manning, Defence Research Board, from near Cape Kellett at the southwest end of the Island (Prest, 1957, p. 457). This analysis indicated an abundance of spruce, pine, birch, and alder pollen. Such an assemblage indicates a climate much milder than the present, which supports only an Arctic type of tundra and bog with dwarfed willow as the largest plant on the tundra. The sediments were therefore considered as probably Aftonian or Yarmouth interglacial deposits. Manning was the first to recognize definite glacial deposits in the western part of the Island (Manning, 1956). A stratigraphic sequence has since been established for the western part of Banks Island by J. G. Fyles (Craig and Fyles, 1960) as follows (from the bottom upwards) :— Beaufort formation; till; lenticular pond silts with peaty materials and locally including small trees and "beaver-sticks"; a bouldery layer that may be either till or colluvium; post-glacial peat, and the present Arctic tundra and bog surface. A radiocarbon dating of a sample from the base of the peat indicates an age of more than 35,000 years (I-GSC-19). Fyles records: "The peat is uncompressed and hence has not been overridden by glacial ice. The radiocarbon age supports the inference

[7] J. Terasmae, personal communication.

that the western part of Banks Island was not glaciated during the Wisconsin." Fyles has, furthermore, mapped the approximate limit of the Wisconsin
glaciers in southern and eastern Banks Island; this is shown on Map 1 as a
single line of dots.[8] Here on western Banks Island, then, is an excellent
record of both older glacial and interglacial deposits. Two other occurrences

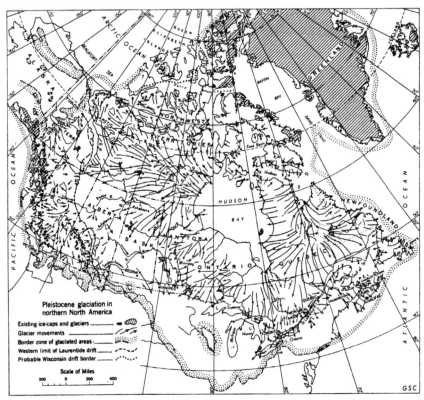

MAP 1. Pleistocene glaciation in northern North America

of buried organic deposits that may represent an interglacial stage are near
Eureka on Ellesmere Island, and on the Stuart River on Bathurst Island
(Prest, 1957, p. 457). The former site was studied by N. J. McMillan and
the latter by D. J. McLaren both while on Geological Survey bedrock
reconnaissance projects.

[8]The glacial border on Banks Island, as shown on the 1958 Glacial Map of Canada,
was drawn by the writer to separate roughly the reported glaciated eastern part of the
Island from the presumably unglaciated western part. It is now obvious that the
reported glacial limit actually represents the Wisconsin limit of glaciation and that it
should have been farther east on Banks Island. As this same situation may well prevail
in the case of other northwestern Q.E. Islands the writer has extended Fyles' Wisconsin
limit to Bathurst Island and northward in place of the glacial limit as shown on the
1958 map.

From the foregoing, it is clear that there is substantial evidence in Canada of early and mid-Pleistocene glacial and interglacial deposits. Most of these deposits are found in the peripheral areas of glaciation where scouring action has not been excessive.

## THE WISCONSIN GLACIAL STAGE

The deposits of the Wisconsin glacial stage comprise by far the greatest area and bulk of Canada's soils, so that some concept of the general sequence of events, no matter how nebulous or controversial, is worthy of attention. As earlier mentioned, the chronology of this stage is at present a subject of keen debate. The crux of this discussion lies in the fact that the generally accepted or standard chronology of the Wisconsin glacial stage began with the Farmdale glacial substage as recognized in southern Illinois, and its age was considered as ranging from 22,000 and 25,000 years B.P. The great gap between the end of the Farmdale substage and the beginning of the Wisconsin, possibly as much as 100,000 years ago, was assumed until recently to have been occupied by glacial processes as the ice built up to its maximum and incorporated ice-frontal debris as it spread outwards. Any deposits older than (say) 25,000 years that might have escaped erosion would therefore have to be Sangamon or Illinoian or older. But over the past five to eight years, radiocarbon datings of more than 30,000 years have been made on numerous deposits that are definitely post-Sangamon, and on similar deposits believed to be so. These have been described from locations in several of the northern United States as well as in Canada. This has led to speculation concerning the recognition and naming of another major glacial stage and of a following interglacial stage between the Sangamon and the "classical" Wisconsin.

A more generally favoured view is, however, that the definition of the Wisconsin should be broadened to include such deposits and the inferred events. In 1957 the writer expressed the opinion that "advance" substages of the Wisconsin as well as "retreat" substages (that began with the Farmdale) should be recognized. This concept is gaining some recognition in both the United States and Canada as older non-glacial deposits of diverse ages are recognized. Recently, an Illinois State Survey report (Frye, J. C. and Willman, 1960) indicates that the Farmdale loess includes much silt that is non-glacial in origin, and that till referred to as Farmdale is not so. The authors conclude that the Farmdale is a major period of glacier withdrawal in the Lake Michigan lobe, rather than an episode of glacier advance. They furthermore outline a revision of the Wisconsin stage applicable to Illinois. This revised terminology is being received favourably in some quarters outside that state.

The accepted sequence and timing of Wisconsin events is rendered even more suspect by undeniable evidence of a major glacier advance from the Lake Erie region beginning about 28,000 B.P., and reaching southern Ohio about 17,000 years B.P. On the north side of the Lake Erie basin, forests were overridden by an advancing glacier about 28,000 years B.P. (L. 185 B) according to Dreimanis (1959, p. 9). This is consistent with the research of R. P. Goldthwait (1958) and fellow workers in Ohio where a well-documented sequence of radiocarbon datings on wood incorporated in till indicates an orderly advance of a major glacier from Lake Erie in a south-westerly and southerly direction. Forests, indicative of a cool, moist, climate, were overridden at Cleveland about 24,500 years ago (W71, 24,600±800), near Columbus 23,000 years ago (Y499, 23,000±850), and at the southern terminus approximately 35 miles north of Cincinnati, less than 17,000 years ago (Y450, 16, 560±230). Thus the record reveals a significant advance of one major glacial lobe while another was at or near its maximum development and had even begun to retreat. The southernmost large moraine in Illinois, the Shelbyville, is dated at 19,200±700 (W-187) and associated loess at 20,700±650 (W-399).

Beneath the till bearing the wood in Ohio there is a buried "soil" or palaeosol, developed on gravels and in one place on till. These soils do not resemble the Sangamon soils in southern Ohio. Goldthwait (1958, p. 213) concludes that these "soils" do not indicate a long interglacial period but "that they represent one interstadial period occurring during 5,000 to 15,000 years between two long major advances of the Wisconsin stage."

The chronology of the Wisconsin has been rendered even more confusing by the dating of some Iowan deposits as older than Farmdale rather than younger. Thus, much detailed evidence and painstaking work in one area stands, seemingly, in direct opposition to that in another. At the moment, although progress is being made, the problem of an acceptable Wisconsin chronology remains unresolved.

If the study of the geology of Canadian soils is to advance, however, there must be a framework to build on, or to adjust, as necessity demands. Some years ago Dreimanis tentatively correlated his preclassical Wisconsin tills on Lake Erie with the Würm I of Europe. The writer concurs and suggests that the terms Wisconsin I and II might well be applied to late Pleistocene deposits in North America. The objection that this is unwise in that some day still another major substage may be recognized does not appear valid, when it is recalled that the pulsating "advance" and halting "retreat" of the last major continental glacier is being considered. Use of the terms Wisconsin I and II is in keeping with the broad binary nature of the last glacial stage in Europe, and with the deductions of Goldthwait (1958) in Ohio. Use of the term Wisconsin I avoids introduction of a new name where it is not necessary, allows for further subdivision as "advance" sub-stages are documented, and provides the logical setting for a mid-Wisconsin "break" prior to the "retreat" substages of Wisconsin II time.

In general, Canadian workers may feel inclined to "side-track" or over-look the Wisconsin chronological debate as not directly applicable to the geology of the soils of Canada. As work progresses in southern Canada, however, it is found that the issue cannot be conveniently side-stepped. In the first place, non-glacial deposits which do not appear to be Sangamon but are too old for the standard Wisconsin have already been recognized.

Some of these are as follows:

1. The St. Pierre beds, and the St. Pierre Interval as recognized by Gadd (1959) in the Three Rivers Region, Quebec. Radiocarbon datings at the Groningen laboratory by H. deVries, employing a process of isotope enrichment, indicate a finite age of about 65,000 years B.P. (Gro. 1711 and 1776), as quoted by Dreimanis (1959, p. 26). As mentioned earlier in this paper, an "advance" Wisconsin interstadial must be considered in connection with these beds.

2. The non-glacial deposits of the James Bay Lowland, Ontario and in particular the Missinaibi beds as described by F. H. McLearn (1926) and by O. L. Hughes, with palynological studies by J. Terasmae (1958). These deposits, as already mentioned, have been placed in the range 55,000 to 65,000 years B.P., and may be "advance" interstadial deposits. (Terasmae and Hughes, in press). In relation to palynology, they may well be the more northerly equivalent of the St. Pierre sediments. Supporting evidence of a Wisconsin age from both the St. Pierre and Missinaibi beds comes from somewhat similar deposits in the extreme southwest of New York State (Muller, 1960, pp. 11 and 25). Buried peat from the Otto site has been dated at Groningen at greater than 52,000 years (Gro. 2565). The plant assemblage is unlike that considered typical of the warmest part of the Sangamon interglacial stage.

3. The Port Talbot interstadial interval of Dreimanis, as indicated by gyttja in bedded silt that is both underlain and overlain by two tills. This is known to be more than 40,000 years B.P. (L. 370A). A peat ball found by Dreimanis has recently been dated at Groningen at 47,000 years B.P.[9]

4. The interval indicated by wood found by Dreimanis in till at Plum Point, Ontario. Radiocarbon datings range from 24,600±1600 to 28,200± 1500 (L 217B and L 185B respectively).

5. The Quadra beds of the Pacific Coast for which Fyles has obtained several datings ranging from 25,000 to more than 40,000 years B.P. Finite dates over this range would appear to indicate a major interstadial period or mid-Wisconsin "break."

6. Numerous other samples from various parts of Canada analysed over the past five to ten years that were beyond the range of the laboratory concerned and which are considered to be probably post-Sangamon.

In the second place the problem of a Wisconsin chronology has to be faced in that recent work indicates that the basic history of the Great Lakes is not yet resolved. Events in the Lake Erie and Lake Ontario basins do not readily correlate with those in the upper Great Lakes. Papers by Dreimanis and Karrow in this Symposium touch on these problems. It is sufficient to say here that the Lake Ontario basin may have been free of ice from Two

[9]A. Dreimanis, personal communication.

Creeks time—about 11,400 years ago—to the present, instead of being reoccupied for a lengthy period after that short but important interstadial period. This concept is in keeping with present-day acceptance of shell dates from the Champlain Sea of the order of 11,000 years in the St. Lawrence and Ottawa River valleys, and in particular with the work of Gadd in the Three Rivers region (Gadd, 1959). Shells from a coquina at an elevation of 542 feet, recognized by the author as the marine limit on Mount Royal, Montreal, have been dated at 11,370±360 (Y-233). This is the oldest date yet obtained, as might well be expected if the radiocarbon determinations are internally correct.

Thirdly, with by far the greater part of Canada deglaciated within a period of only 10,000 to 15,000 years, it appears essential that Canadian workers should take an active part in the establishment of a late Wisconsin chronology. Yet the accepted chronology of American associates must be used as a base and endeavour made to reach mutual understanding regarding the early stages of the Wisconsin glacier retreat.

Finally, the close of the last glacial stage and the events of the Recent non-glacial stage are of great interest in Canada. This interval probably occupies the last 7,000 years, though many workers favour a date of only 5,000 years based on stabilization of sea-level. The date of 7,000 years B.P. for the start of the Recent interval is strengthened by work in both the Keewatin and Labrador–New Quebec regions indicating that the continental glaciers disappeared at about this time (Craig and Fyles, 1960, pp. 10, 11).

## SOILS

From the foregoing it may perhaps be evident that geologists, assigned to the task of compiling a broad regional map of the soils of Canada, might well start off by distinguishing four main classes, dependent upon the geological history of the areas concerned: (1) soils of the unglaciated areas; (2) soils from areas glaciated prior to the Wisconsin stage; (3) soils from areas glaciated during the Wisconsin; and (4) soils of the Recent non-glacial interval. The third class and the most important, might then conveniently be subdivided according to the origin of the deposit.

1. In those parts of the unglaciated areas where erosion has not been excessive, or at least where it has lagged behind weathering processes, and where permafrost conditions have not always prevailed, mature soil profiles might be expected. Such areas in Canada are of limited extent or have been little studied. The largest area is in western Yukon Territory. The southern part of this region is the most important from the standpoint of future agronomy. Another large area of this type may be present in the north-western Queen Elizabeth Islands. There may be two unglaciated areas in the Foothills west of the Mackenzie River opposite Great Slave and Great Bear Lakes. In all these cases, however, the soils have been little studied by either geologists or pedologists. A small unglaciated area of considerable

interest lies in the Foothills of southwestern Alberta; it comprises some excellent ranch lands. The contrasting physiography of this area and the surrounding glaciated region is readily apparent from the air, on flights passing between Lethbridge and Vancouver. Detailed pedological studies in and around this area would be of great interest to geologists. Other very small unglaciated areas in Canada occur on the International Boundary west of Milk River, in the western part of the Cypress Hills, and projecting northward from the States into south-central Saskatchewan at longitude 106°. None of these areas have been adequately studied either geologically or pedologically.

2. Areas in Canada that were glaciated only prior to the Wisconsin or last glaciation are indeed limited. Geological Survey work in the Yukon has indicated some areas where the older glaciations reached somewhat higher and spread more laterally than the Wisconsin glaciers.[10] This phenomenon is of considerable importance in connection with the placer mining industry. In other places in the Cordillera too, an older Pleistocene glacier reached higher than the Wisconsin glacier. Recent Geological Survey work by Fyles on Banks Island has indicated that most of the western part of the island falls into this class, and the same may apply to some adjacent areas (Craig and Fyles, 1960). In eastern Canada only the Magdalen Islands appear to have escaped direct glaciation in Wisconsin time, although floating glacier ice reached their shores and deposited considerable stratified drift and minor water-deposited till (glacio-natant till) (Prest, 1957).

3. The third class of soil in the geological-historical sense is that representing Wisconsin glaciation; it comprises most of Canada's land surface. Even in areas of abundant rock outcrop, as indicated on the insert map in the Glacial Map of Canada 1958, there is a variable mantle of till or other drift on the outcrops or between them; this mantle supports a surprising amount of vegetation and attendant animal life. These Wisconsin-age deposits have been receiving much attention from geologists, pedologists, geographers, biologists, archaeologists, and others. Wisconsin glaciation has given Canada its soils; these parent materials may readily be subdivided according to their mode of origin.

The glaciers themselves blanketed great tracts of land with an unsorted mixture of materials commonly known as till. This mantle is usually referred to as ground moraine. The till is generally a mixture of soil materials from clay-size to boulders, but the percentages of any one size vary greatly from place to place, depending on the nature of the terrain that has been glaciated. In and around the Canadian Shield, for example, the till is usually sandy or gravelly and is in fact often mistaken for a gravel. Yet even within the Shield the till may be clayey, as is the case with the Cochrane till in northern Ontario where the glacier readvanced from the Palaeozoic limestone areas of the James Bay Lowland and overrode the varved clay terrain of glacial Lake Barlow–Ojibway. This till can be distinguished in many

[10]H. S. Bostock, personal communication.

places only with great difficulty from massive beds of lake-deposited clay. In southern Ontario and the St. Lawrence River valley, where the glaciers have overridden tracts of Palaeozoic limestone and in some cases minor glacial lake basins, the tills are again very clayey or silty. Due to the common admixture of stones of all kinds, sizes and shapes they are very tough materials and cause considerable trouble during major construction work, although in general they form good foundations.

In western Canada, west of the Canadian Shield, the ground moraine till is also very clayey. The last glacier advanced over older clay-rich tills that had derived much of their fine material from the underlying Palaeozoic and Mesozoic limestones and shales; in places the last glacier scoured these rocks afresh. As the Mesozoic shales are clay-rich, and also in places contain appreciable bentonite, the clayey tills (and younger lake clays) have swelling and shrinking properties that are most troublesome from the engineering point of view. These same Wisconsin tills, however, underlie a large percentage of the great western farm lands.

In the Cordillera the complex geology is reflected in the variability of the tills: they vary from clayey to gravelly. The tills of the Arctic Islands are probably also considerably varied, but here the influence of solifluction is so great that the true nature of the tills is but little known. In some places large tracts of land are surfaced with a mantle of solifluxed, sticky, clayey material, whereas in others the terrain is mantled with frost-heaved bedrock, referred to as "felsenmeer."

Aside from ground moraine, appreciable tracts of land are covered by end moraine, and hummocky moraine. These, in the main, are broadly similar to associated ground moraine in composition although generally more sandy or gravelly; they may include stratified drift. In eastern Canada, end moraines and interlobate moraines are usually less than ten miles wide but many tens of miles long. They may form prominent features rising several hundred feet above the surrounding country and they provide excellent tourist resort areas. End moraines mark ice-frontal positions of active glaciers. In western Canada, end moraines are generally more subdued than in the east, and are often traceable only with difficulty. The prominent morainal terrain in the west is the wide-scale "hummocky" type that indicates only semi-active ice or general stagnation over large areas. Such moraines may have a relief of a few hundred feet and provide both resort areas and localized tracts of tillable land.

Glacio-fluvial deposits are also widespread in Canada. The Glacial Map of Canada, 1958, indicates the prevalence of large eskers over the Shield areas, and this reflects the general character of the bedrock. But eskers are not uncommon elsewhere: the Glacial Map really only shows the eskers where air photo studies have been made. Meltwater channels and areas of outwash are similarly more widespread in Canada than the map indicates, but work has not progressed to the point where these have been differentiated for mapping purposes.

The Wisconsin glaciers were also responsible for great tracts of land being covered by vast glacial lakes or being overlapped by the sea. The maximum area covered by these waters is shown on Map 2. This map does not indicate the size of any one lake at any one phase, but shows a gradually receding system of lakes as the ice margin retreated and new outlets were opened. In the case of the maximum marine overlap, however, the sea covered the

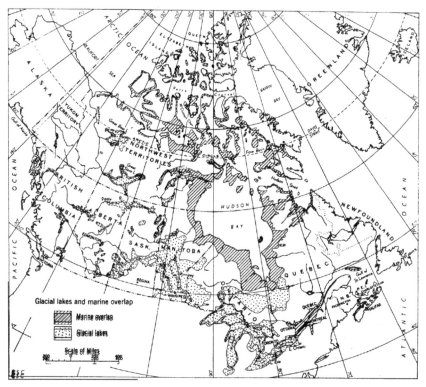

MAP 2. Glacial lakes and marine overlap

entire land area indicated and receded to approximately its present position as the land areas slowly rebounded, following the retreat of the glaciers. In some areas the land is still actively rising whereas in others low shores have been or are being submerged in accord with eustatic or world-wide sea level changes.

Finally there is the fourth class of soils of Canada—Recent stage deposits and the existing ice caps and glaciers. Perhaps the only areally significant deposits of alluvium observable today are in the Fraser and Mackenzie River deltas; and in the latter case they are but little known. Much of the low-lying sea coasts have no doubt emerged during Recent time but these cannot be conveniently separated from those of late Wisconsin emergence. Thus the present-day glaciers and ice caps remain as the best expression of the Recent "soils" of Canada. These consist of ice with much incorporated debris yet to be deposited. Although some glaciers are known to have advanced in recent years the major glaciers and ice caps appear to be receding. It is conceivable that if the present non-glacial period, the Recent stage, continues for a few thousand years more, much new soil will become available for study, although some lowland areas will be partially submerged by the resulting rise in sea level. Some mountain glaciers may expand according to the delicate balance between available moisture, height of land and mean annual temperature. It should be kept in mind that the climate of today is generally cooler than that of 5,000 years ago, which is referred to as the climatic optimum. No one can say whether the climate will warm up to this point again or whether the Recent non-glacial stage will give place to another glacial stage.

## REFERENCES

1. Armstrong, J. E. (1960). Surficial geology of New Westminster map-area, British Columbia. Geol. Surv., Can., Paper 57-5.
2. Bostock, H. S. (1936). Carmacks district, Yukon. Geol. Surv., Can., Mem. 189.
3. ———— (In press). McQueston map-area. Geol. Surv., Can.
4. Christiansen, E. A. (1960). Geology and groundwater resources, Qu'Appelle area, Saskatchewan. Sask. Res. Council, Geol. Div., Rept. 1.
5. Craig, B. G., and Fyles, J. G. (1960). Pleistocene geology of Arctic Canada. Geol. Surv., Can., Paper 60-10.
6. Dreimanis, A. (1959). Friends of Pleistocene geology, Eastern Section. Guidebook, Contrib. no. 25, Dept. Geol., Univ. Western Ontario.
7. Flint, R. F. (1957). Glacial and Pleistocene geology. New York: John Wiley and Sons.
8. Frye, J. C., and Willman, H. B. (1960). Classification of the Wisconsin stage in the Lake Michigan glacial lobe. Illinois State Geol. Surv., Circ. 285.
9. Fyles, J. G. (in press). Surficial geology of the Horne Lake–Parksville map-areas. Geol. Surv., Can., Mem. 318.
10. Gadd, N. R. (1959). Surficial geology of the Bécancour map-area, Quebec. Geol. Surv., Can., Paper 59–8.
11. Goldthwait, R. P. (1958). Wisconsin age forests in western Ohio. Part I, Ohio Jour. Sci., pp. 209–219.
12. Logan, Sir William (1847). Report of progress for the year 1845–6, pp. 71–75. Geol. Surv., Can.
13. Manning, T. H. (1956). Narrative of a second Defence Research Board expedition to Banks Island, with notes on the country and its history. Arctic, 9: nos. 1 and 2; Appendix II (Pollen Analysis by J. Terasmae, pp. 64–65).
14. McLearn, F. H. (1926). The Mesozoic and Pleistocene deposits of the Lower Missinaibi, Opasatika and Mattagami rivers, Ontario. Geol. Surv. Canada, Sum. Rept., 1926, Pt. C, pp. 16–44.
15. Muller, E. H. (1960). Friends of the Pleistocene, Eastern Section. Guidebook; Glacial geology of Cattaraugus County, N.Y. Contrib. Dept. Geol., Syracuse University.
16. Prest, V. K. (1957). Pleistocene geology and surficial deposits. Ch. vii in Geology and Economic Minerals of Canada, ed. C. H. Stockwell, 4th ed.; Geol. Surv., Can.
17. Terasmae, J. (1959). Palaeobotanical study of buried peat from the Mackenzie River delta area, Northwest Territories. Can. Jour. Bot., 37: pp. 715–17.

18. ——— (1958). Contribution to Canadian palynology. Geol. Surv., Can., Bull. 46.
19. TERASMAE, J., and HUGHES, O. L. (1960). A palynological and geological study of Pleistocene deposits in the James Bay Lowlands, Ontario. Geol. Surv., Can., Bull. 62.
20. TYRRELL, J. B. (1891). Pleistocene of the Winnipeg Basin. Amer. Geol., 8: pp. 19–28.

ADDENDUM 1965

FIELD INVESTIGATIONS of the Cape Breton Island subtill organic deposits by the writer in 1964, followed by palynological studies and radiocarbon datings in the laboratories of the Geological Survey of Canada, have resulted in a better picture of the stratigraphic sequences and the implied history. At Bay St. Lawrence unconsolidated sediments form a sea-cliff up to 150 feet in height; most of the upper part of the section is a gravelly till while near the bottom there are stratified sediments, including an organic layer, overlying a basal till. The stratified sediments are believed to represent a flood plain deposit. As a radiocarbon dating on a piece of the buried wood indicates an age of more than 38,270 years (GSC-283) and palynological studies indicate a climate considerably cooler than the present, it may be inferred that an early Wisconsin non-glacial interval rather than an inter-glacial interval is represented by the organic materials.

On the Mabou River near Hillsborough in southwestern Cape Breton Island, up to 12 feet of a clayey till overlies some 6 feet of stratified clay and silty clay containing lenses of organic debris and with a woody basal layer or up to 4 feet of compressed peat. The organic-bearing sediment locally overlies a few inches to a few feet of yellowish silty clay that in turn rests on a thin ortstein layer developed on the irregular surface of highly oxidized sand grading downward into oxidized sandy gravels, the whole comprising some 10 to 12 feet. The base of the exposed section is at road level, about 11 feet above the river, here only a few feet above sea-level. The radiocarbon age is more than 38,000 years, but palynological studies suggest a climate cooler than the present; hence, again, the non-glacial interval may be early Wisconsin.

Approximately 30 miles east of Hillsborough, in the village of Whycoco-magh on the Bras D'Or Lakes, plant-bearing beds between gravelly till layers are again present close to sea-level. The pollen assemblages are similar to those near Hillsborough, and correlation is further substantiated by a new radiocarbon dating, on wood from the stratified beds, of > 44,000 years (GSC-290). Further eastward, at Benacadie, several feet of well-bedded silts lying between an upper 50 feet of stony clay-silt till and a lower 20 feet of gravelly till contain a pollen assemblage somewhat similar to part of the Bay St. Lawrence section.

Buried organic deposits on the Inhabitants River in extreme southwestern Cape Breton Island and at Milford in central mainland Nova Scotia (Prest, 1957) were not examined in 1964; they may represent either 'advance' Wisconsin or pre-Wisconsin intervals. Wood from the Milford site has been dated as > 33,800 years (GSC-33).

# SOILS OF THE COASTAL AREA
# OF SOUTHWEST BRITISH COLUMBIA*

## J. E. Armstrong, F.R.S.C.

COMMENCING IN 1949 the writer undertook for the Geological Survey of Canada a geological investigation of the Fra er Lowland of British Columbia. In the course of this study he co-operated extensively with consulting soils engineers and pedologists and, as a result, observed the need of both groups for basic geological information, especially in view of the complexity of the Pleistocene and Recent stratigraphy and history of this area.

### PHYSICAL FEATURES

The Fraser Lowland is a triangular-shaped area of low relief with its apex sixty-five miles east of the Strait of Georgia (see Figure 1). It is

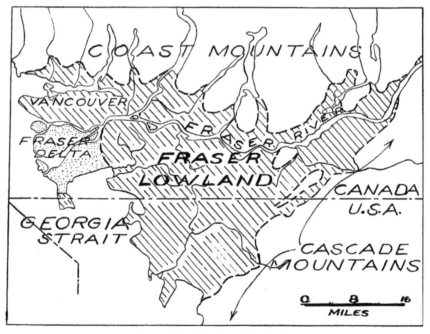

FIGURE 1.    The Fraser Lowland

*Published by permission of the Director. Geological Survey of Canada.

bounded on the north by the Coast Mountains, on the southeast by the Cascade and Chuckanut Mountains, and has an area of approximately 1,350 square miles. The Canada–United States boundary divides the lowland along the 49th parallel and as a result 390 square miles are south of the boundary and 960 square miles north. The writer has studied in detail only that area north of the boundary.

The Coast Mountains rise abruptly to 4,000 to 7,000 feet above deep U-shaped valleys, which are occupied by rivers, lakes, and arms of the sea. The Cascade Mountains rise to 7,000 feet or more. The Chuckanut Mountains are a lower front range of the Cascades. Chilliwack, Sumas and Vedder Mountains are isolated mountain areas to the northwest of the Cascade Mountains, separated from the main mountain mass by wide flat-bottomed valleys.

The dominant topographic feature of the Fraser Lowland is the Fraser River which occupies a post-glacial valley up to three miles wide and fifty feet or more deep. It terminates in a growing delta nineteen miles long and fifteen miles wide. North and south of the Fraser River the Fraser Lowland consists mainly of wide, flat-topped, and gently rolling, low hills separated by wide flat-bottomed valleys. Most of these hills consist largely of unconsolidated deposits and do not exceed 500 feet in height, although three bedrock hills exceed 1,000 feet. The hills range in area from 1 to 150 square miles.

## GEOLOGY

The Fraser Lowland was subjected to at least four glaciations—three probably major and separated by intertill or interglacial intervals, and one probably valley glaciation only. During each major glaciation the land was depressed relatively to the sea, possibly as much as 1,000 feet. As a result of this complex geological history very thick deposits (1,000 to 3,000 feet) of widely diversified origin were laid down. Geological events in Pleistocene and Recent time in the Fraser Lowland are synopsized in Table I. The oldest event is number 1 and the youngest number 10.

TABLE I
SEQUENCE OF EVENTS IN PLEISTOCENE AND RECENT TIME, FRASER LOWLAND

10. Salish Group (Post-Glacial deposits still being formed; in part overlap Capilano deposits): Alluvial, deltaic, estuarine, marine, lacustrine, colluvial, and swamp deposits. Includes Fraser River delta deposits at least 700 feet thick. Salish deposits are believed to have been laid down after the present relative position of the land and sea first became established. Some evidence has been found indicating that conditions have not remained stable and that one or more submergences and emergences of as much as fifty feet have taken place during deposition of Salish deposits. Radiocarbon dating indicates Indian middens in the Fraser Lowland, at or near sea level, to be up to 2900 ± 170 years old. Forty miles east of the lowland at an elevation of approximately 225 feet, 50 feet above the Fraser River, a midden 8150 ± 310 years old has been found.

9. Capilano Group (Post-Vashon non-glacial deposits no longer being formed. In part older, in part contemporaneous, and in part younger than Sumas group): Marine off-shore, littoral, and delta deposits, alluvial and aeolian deposits formed during uplift of the land above the sea following the retreat of the Vashon ice-sheet. This emergence, which exceeded 600 feet, was not uniform. Evidence suggests that before the advance of Sumas ice, the land emerged to within 100 feet or less of its present elevation and then was submerged at least 500 feet, emerging during the wasting of Sumas ice. The off-shore marine deposits included here have a maximum thickness of at least 900 feet.

8. Sumas Group (Post-Vashon glacial deposits related to valley ice): This glacier terminated about 25 miles east of Vancouver. The glacier responsible for these deposits advanced into the sea depositing glacio-marine sediments (marine drift) up to 300 feet thick. As the land rose the glacier was grounded, depositing till, and as the glacier wasted, recessional outwash and glacio-lacustrine deposits up to 450 feet thick were laid down. Wood collected from the glacio-marine sediments in five localities indicates the ice advance commenced not more than 11,500 years ago. This glacier probably is Valders in age.

7. Vashon Group (Deposits of last glaciation of continental ice-sheet proportions): As the ice-sheet advanced, outwash up to 200 feet thick was deposited in front of the glacier and till beneath it. During the ice advance the land was depressed at least 750 feet below present sea level and during wasting of the ice-sheet the glacier thinned, losing contact with the sea floor and became floating ice from which debris was dropped to form marine drift. Shells collected in two localities gave radiocarbon dates of approximately 12,500 years. In Washington State peat overlying Vashon till is in places 14,000 ± 900 years old. The Vashon glacier may possibly be Tazewell in age.

6. Erosion Interval: During this interval considerable relief was developed on the underlying deposits and as a result Semiamu deposits have been eroded away in much of the area.

5. Semiamu Group (Deposits related to glaciation): Till and stratified drift very poorly exposed and probably missing in much of the area as explained above. The glacier responsible for these deposits was probably of continental ice-sheet proportions.

4. Erosion Interval: During this interval considerable relief was developed on the underlying deposits.

3. Quadra Group (Intertill, probably interglacial sediments): Alluvial sediments deposited during a time which had a temperate climate somewhat similar to present-day climate. Included here is Point Grey Beds exposed in University sea cliffs and containing peat which, according to tentative radiocarbon dating, is 30,000 to 40,000 years old. Quadra sediments on the east coast of Vancouver Island have yielded radiocarbon dates that vary between 25,000 and 40,000 years.

2. Seymour Group (Deposits related to glaciation): Tills, and associated glacio-lacustrine, glacio-fluvial, and glacio-marine deposits laid down during several local advances and retreats of a Cordilleran ice-sheet. Maximum thickness 500 feet.

1. Pre-Seymour Group: Older glacial, intertill or interglacial and pre-glacial sediments as indicated by deep well records. Insufficient evidence available to determine actual sequence, and some Pliocene may be included here. Maximum thickness 1,500 feet.

## Nature of Deposits

The nature of the deposits is synopsized in Table II.

TABLE II

NATURE OF PLEISTOCENE AND RECENT DEPOSITS

10. Salish Group:
   Alluvial deposits; silt, silty clay, organic clay, sand, and gravel
   Deltaic deposits; sand and gravel
   Estuarine and deltaic deposits (Fraser River delta); sand, silt, silty clay, and clayey silt
   Marine deposits (shore); sand and gravel
   Lacustrine deposits; sand, silty sand, clayey silt, and silty clay
   Colluvial deposits; gravel, sand, silt and talus
   Swamp deposits; peat
9. Capilano Group:
   Marine deposits (off-shore); silty clay, clayey silt, silt, sand, and minor gravel; poorly sorted till-like mixtures
   Marine deposits (shore); deltaic, sand and gravel; beach sand and gravel; littoral sand
   Alluvial deposits; gravel and sand
   Windblown deposits; sand
8. Sumas Group:
   Glacial deposits; till
   Glacio-fluvial deposits; outwash sand and gravel
   Glacio-lacustrine deposits; silt, clayey silt, silty clay, clay, minor sand and gravel
   Glacio-marine deposits; stony clayey silt, stony silty clay, stony sandy silt, silty clay, clayey silt, silt; minor sand and gravel
7. Vashon Group:
   Glacial deposits; sandy to silty till and substratified drift
   Glacio-fluvial deposits; outwash sand and gravel
   Glacio-marine deposits; stony clayey silt, stony silty clay, poorly sorted till-like mixtures; minor clayey silt, silty clay, silt, sand, and gravel
5. Semiamu Group:
   Glacial deposits; sandy to silty till and substratified drift
   Glacio-fluvial deposits; outwash sand and gravel
   Glacio-lacustrine deposits; varve-like silt and clay
3. Quadra Group:
   Alluvial deposits, sand, gravel, silt, and minor silty clay, organic clay and peat
2. Seymour Group:
   Glacial deposits; silty to clayey till and substratified drift
   Glacio-fluvial deposits; sand and gravel
   Glacio-lacustrine deposits; varve-like silt and clay, sand
1. Pre-Seymour Group:
   Clay, silty clay, clayey silt, silt, sandy silt, gravel, till, and glacio-marine deposits

The terms clay, silt, and sand as used in this paper are based on the diameter of the constituent particles and are used as follows: clay, less than 0.002 mm.; silt, 0.002 to 0.05 mm.; and sand, 0.05 to 2 mm. The term gravel is used to refer to any stratified deposit in which fifty per cent or more of the constituent particles are larger than sand. No detailed mineralogical studies have been made on the clays and silts. They are believed to consist in a large part of feldspar and quartz and other common rock minerals.

Recent work at the University of British Columbia indicates, however, that they contain some clay minerals. The sands consist largely of quartz and feldspar with lesser amounts of ferro-magnesian and opaque minerals and rock fragments, the proportions varying from place to place. The varved silts and clays are glacial lake deposits consisting of alternating light and dark coloured layers from a fraction of an inch to several inches thick.

Glacio-marine and associated marine deposits containing isolated sea shells are abundantly represented in the Sumas and Vashon groups. They are massive to stratified, soft plastic materials, with silty clay, clayey silt, and silty sand textures. Many contain no stones and very little sand and are normal sea bottom muds. Others, here called stony clayey silt, stony silty clay, stony sandy silt, and related till-like mixtures, contain a small percentage of pebbles, cobbles, and boulders. Some of these appear to have been formed by submarine slumping or by reworking of glacial tills by tides and waves. Most of them, however, are believed to be of glacio-marine origin and to have resulted from the mixing of debris carried by floating glacial ice with sea-bottom mud. Mechanical analyses of representative samples of glacio-marine deposits are given in Table III.

The glacial tills in the area are compact, unsorted mixtures of sand, silt, clay, pebbles, cobbles, and boulders deposited directly beneath the ice. Mechanical analyses of representative samples of tills are also given in Table III.

TABLE III

MECHANICAL ANALYSES OF REPRESENTATIVE SAMPLES OF TILLS AND GLACIO-MARINE DEPOSITS (FINE FRACTION ONLY)

| Locality | Group and Nature of Deposit | Mechanical Analysis | | |
|---|---|---|---|---|
| | | % Clay | % Silt | % Sand |
| | Seymour Group | | | |
| 1. C.P.R. tracks at Ioco Pipeline | Till | 9.2 | 49.5 | 44.3 |
| | Semiamu Group | | | |
| 2. G.N.R. tracks, Ocean Park | Till | 13.7 | 41.7 | 44.6 |
| 3. English Bluff, Boundary Bay | Till | 6.8 | 51.4 | 41.8 |
| 4. Brunette Creek, North Road | Till | 5.5 | 33.7 | 61.8 |
| | Vashon Group | | | |
| 5. Ioco Pipeline, 400' elevation Burnaby Mountain | Till | <.1 | 29.8 | 70.2 |
| 6. Boundary Bay Gravel Pit | Till | <.1 | 43.6 | 56.4 |
| 7. Boundary Bay Gravel Pit | Till | 4.0 | 43.6 | 52.4 |
| 8. Delta Gravel Pit, West of Scott Road | Till | 3.7 | 50.7 | 45.6 |
| 9. Capilano Damsite | Till | 4.4 | 43.2 | 52.4 |
| 10. Lougheed H'wy. Gravel Pit near Essondale | Till | 3.5 | 42.5 | 54.0 |
| 11. Gravel Pit, North Road | Till | 3.7 | 43.1 | 53.2 |
| 12. Gravel Pit, Austin Road | Till | 1.0 | 30.4 | 68.6 |
| 13. G.N.R. tracks, Ocean Park | Till | 3.4 | 33.5 | 63.1 |
| 14. Boundary Bay Gravel Pit | Glacio-marine | 6.6 | 34.6 | 58.8 |
| 15. G.N.R. tracks at Sapperton | Glacio-marine | 11.6 | 52.4 | 36.0 |

| | | | | |
|---|---|---|---|---|
| 16. | Gravel Pit, North Road | Glacio-marine | 9.2 | 55.8 | 35.0 |
| 17. | Delta Gravel Pit, west of Scott Road | Glacio-marine | 13.9 | 67.5 | 18.6 |
| 18. | King George Highway, near Sunnyside | Glacio-marine | 9.1 | 38.1 | 52.8 |
| 19. | Trans Mountain Pipeline, Jackman Road | Glacio-marine | 1.3 | 33.9 | 64.8 |

Sumas Group

| | | | | | |
|---|---|---|---|---|---|
| 20. | Huntingdon Gravel Pit | Till | 4.3 | 35.7 | 60.0 |
| 21. | Gravel Pit south of Abbotsford | Till | 1.9 | 29.7 | 68.4 |
| 22. | Gravel Pit south of Clearbrook | Till | 7.6 | 57.7 | 34.7 |
| 23. | Vedder Mountain Road | Till | 4.4 | 35.0 | 60.6 |
| 24. | Trans Mountain Pipeline, Jackman Road (above and separated from 19 by gravel) | Glacio-marine | 26.6 | 60.2 | 13.2 |
| 25. | Mount Lehman Road, one mile north of Trans Canada Highway | Glacio-marine | 13.7 | 73.3 | 13.0 |
| 26. | Gravel Pit south of Abbotsford | Glacio-marine | 17.8 | 58.4 | 23.8 |
| 27. | Murrayville Cemetery | Glacio-marine | | | |
| | 1' depth | | 18.0 | 69.0 | 13.0 |
| | 7' depth | | 24.0 | 71.0 | 5.0 |
| | 12' depth | | 28.5 | 52.5 | 20.0 |
| | 18' depth | | 36.5 | 47.5 | 16.0 |
| 28. | Jackman Road, North of B.C.E. RR. | Glacio-marine | | | |
| | 1½' depth | | 4.0 | 65.0 | 31.0 |
| | 6' depth | | 20.0 | 73.5 | 6.5 |
| | 10' depth | | 28.4 | 48.3 | 14.3 |
| | 11½' depth | | 2.5 | 30.5 | 67.0 |
| 29. | Lehman Road between Trans Canada Highway and B.C.E. RR. | Glacio-marine | | | |
| | 1' depth | | 16.5 | 49.2 | 34.3 |
| | 3' depth | | 21.5 | 51.2 | 27.3 |
| | 6' depth | | 29.0 | 51.5 | 19.5 |
| | 12' depth | | 28.0 | 52.0 | 20.0 |
| | 18' depth | | 39.0 | 48.0 | 13.0 |

Capilano Group

| | | | | | |
|---|---|---|---|---|---|
| 30. | Haney Brickyard | Marine | | | |
| | 7' depth | | 46.0 | 47.0 | 7.0 |
| | 17' depth | | 44.0 | 52.0 | 4.0 |
| | 27' depth | | 50.0 | 45.0 | 5.0 |
| | 37' depth | | 46.0 | 47.0 | 7.0 |

*Mechanical analyses 1 to 26 were made by the Geological Survey of Canada. Mechanical analyses 27 to 30 were made by N. Ahmad, graduate student at University of British Columbia.

ENGINEERING GEOLOGY

Adequate information on the kind and distribution of geological materials aid the engineer, planner, and contractor in solving many problems pertaining to foundation materials, drainage, sewage disposal, flood control, slides and washouts, and construction materials. Only the first two aspects will be discussed in this paper.

Of particular interest is the fact that the Vashon group till and older deposits have been preloaded by up to 7,500 feet of ice, whereas the post-Vashon group till deposits west of the maximum advance of the Sumas (14) ice have only been preloaded by the weight of the sediments above them. During the greater part of its life, the Sumas group ice-sheet was apparently floating in the sea. Only in its final stages was it resting on land; consequently sediments underlying this ice were probably subject to a load consisting of only a few hundreds of feet of ice. At its point of maximum advance, the Sumas glacier lay about twenty-five miles east of Vancouver.

The tills should remain relatively undeformed under heavy loads, whereas till-like glacio-marine deposits that have not been preloaded by ice may fail when subjected to relatively light loads. The very different reaction of similar materials to loads is readily explainable when the origins of the two are considered. The tills were deposited under a great weight of ice, whereas the glacio-marine deposits were dropped from floating ice. In the Fraser Lowland, the glacio-marine deposits attain great thicknesses; when major industrial expansion is undertaken in areas underlain by them they will certainly present bearing problems quite dissimilar to those of the tills that they resemble superficially.

Table IV summarizes the characteristics pertinent to foundations of the various materials found in the Fraser Lowland.

TABLE IV

FOUNDATION MATERIALS AND DRAINAGE*

| Type of Deposit | Value for Foundations | Drainage |
|---|---|---|
| Tills of Vashon, Semiamu, and Seymour Groups | Excellent bearing value, have been preloaded once or more by several thousand feet of glacier ice | Surface drainage only |
| Sumas Group till | Good to excellent bearing value, has been preloaded by at least several hundred feet of glacier ice | Mainly surface drainage, but also limited downward drainage |
| Sand and gravel of Semiamu, Quadra, and Seymour Groups<br><br>Sand and gravel of Capilano and Vashon Groups where overlain by till | Good to excellent bearing value, have been preloaded by glacier ice | Excellent downward drainage |
| Sand and gravel of Capilano and Vashon Groups where no preloading by glacier ice has taken place<br><br>Sand and gravel of Sumas and Salish groups | Good bearing value | Excellent downward drainage except where ground-water table is within a few feet of the surface |

*Only deposits likely to be encountered are included here.

| | | |
|---|---|---|
| Glacio-marine stony clayey silt, stony silty clay, stony silty sand, silty clay, clayey silt, silt, and sand of Vashon and Sumas Groups where they have been pre-loaded by glacier ice<br><br>Varve-like clay and silt of Seymour and Semiamu Groups | Fair to good bearing values | Surface drainage only |
| Glacio-marine stony clayey silt, stony silty clay, stony silty sand, silty clay, clayey silt, silt and sand of Vashon and Sumas Groups where no preloading by glacier ice has taken place<br><br>Marine silty clay, clayey silt, and silt of Capilano Group<br><br>Glacio-lacustrine silt, clayey silt, silty clay, and clay of Sumas Group | Poor to fair bearing values, may have excessive settlements | Surface drainage only |
| Sand of Salish Group in low flat-lying areas | Fair to good bearing values depending on density | Drainage poor much of the year as ground-water table is within a few feet of the surface |
| Clayey silt, silty clay, sandy silt, silty sand, clay, silt, and fine sand of Salish Group, mostly in low flat-lying areas | Very poor to fair bearing values; in areas where clayey silt, silty clay, and clay predominate may have excessive settle-ments. In areas where sandy silt, silty sand, silt, and fine sand pre-dominate they are susceptible to liquefaction | Surface drainage only |
| Silty clay, organic clay, and peat of Quadra Group | Poor to fair bearing values, have been pre-loaded by several thousand feet of glacier ice | Surface drainage only |
| Peat, clayey peat, peaty clay, and organic clay of Salish Group | Remove from foundations | Very difficult to drain as peat will hold up to 26 times its own weight of water |

Although the tills contain relatively little clay (see Table III), and a high percentage of sand, their coherent nature is partly a result of the weight of glacier ice beneath which they were deposited and is partly a result of their mechanical composition—that is, fine particles fill voids between coarse particles and bind them together to form a natural physical concrete. This feature explains why the tills reconsolidate themselves after sliding, or after excavation when used as fill.

## Agricultural Geology

Modern agricultural soil classification is based upon the nature of the soil profile, which reflects the influence of various factors of soil development including parent material, climate, organisms, topography, time, and geological environment. Parent material and topography are elements in geological environment, but in this paper the term "environment" is used to refer to geological features outside the soil body itself but which may directly or indirectly affect the development of the soil. For example, the stratigraphy and geological structure in and around a particular soil may be an important factor in soil development. This is illustrated by the varieties of soil on the same parent material that have resulted from different kinds of deposits underlying the parent material and thereby setting up different soil development patterns. Drainage generally is recognized as highly important in soil development, but drainage depends mostly on the permeability of the underlying material, which is in turn dependent on the stratigraphy and the depth to a saturated zone. This latter is commonly a function of regional geomorphology and geologic structure. Each of the soil factors mentioned at the start of this paragraph is in itself directly or indirectly dependent on the geological history of the area. Parent material, topography, and geological environment (as defined above) are directly dependent on the geological history, and climate and organisms are indirectly dependent. Parent material, climate and organisms primarily control the kind of weathering. Topography, time and geological environment control the degree of weathering.

The geologist can best help the soil scientist in his interpretation of soils by indicating the role played by parent materials and by geological environment. The soil profiles in the Fraser Lowland, for example, are poorly developed and the texture and composition of the parent material is still dominant. The writer believes that, when the agricultural soils of this area are remapped, the broad divisions of the completed soil map will show a very marked similarity to the surficial geological maps prepared by the writer (Armstrong, 1956, 1957, and 1960).

The moisture content of a soil depends partly on the slope on which it rests, partly on the permeability of the soil profile itself and of the underlying parent material, and partly on the presence or absence of a permanent water-table. The soils of the region fall naturally into two classes based on the downward drainage (permeability) namely—those with restricted downward drainage, and those with good or adequate downward drainage. The soils with restricted downward drainage may be subdivided into those with good surface drainage and those with poor surface drainage.

Soils developed on the following parent materials have restricted downward drainage: till, glacio-marine stony clayey silt, stony silty clay, stony sandy silt and related till-like mixtures; glacio-lacustrine clay and silt; lacustrine clay and silt; alluvial clay and silt; marine clayey silt, silty clay, sandy silt, and silt; organic clay; and peat. All are completely or relatively impervious. The surface drainage on these soils is controlled by local topography. If therefore they occur on a slope, the surface drainage is good, but if they occupy hollows or flat-lying areas, the surface drainage is poor. The areas underlain by till are mainly hilly, and have excellent surface drainage, except locally in depressions and flat areas. The stony glacio-marine deposits have normally a rolling to relatively flat topography; as a result they have only limited areas with good surface drainage. The glacio-lacustrine deposits are mainly exposed on steep slopes and have relatively good surface drainage. All the remaining deposits with restricted downward drainage underlie flat-lying areas and have poor surface drainage.

Sand and gravel of all origins normally have good downward drainage. Exceptions are found in low-lying areas where the water-table is near the surface, permitting little downward circulation of water. Most sand and gravel deposits have relatively flat surfaces, such as outwash plains and valley bottoms.

### AREAS OUTSIDE THE FRASER LOWLAND

The east coast of Vancouver Island is the only other part of southwest B.C. where the surficial deposits have been studied in detail (Fyles, J. G., 1956). Here also unconsolidated deposits are widespread but are generally much thinner than in the Fraser Lowland. Two drift sheets, probably the equivalents of the Vashon and Seymour, are exposed with intertill sediments, probably the equivalent of the Quadra, between them. Except for the absence of deposits similar to those related to the Fraser River, Capilano and Salish sediments are also widespread. With the exceptions noted above, most of the soils found in the Fraser Lowland have equivalents on Vancouver Island with the same engineering and agricultural soil problems.

Similar Pleistocene and Recent deposits to those of the east coast of Vancouver Island are also extensive in the Sechelt and Powell River areas of the Mainland. Most of the remainder of southwest British Columbia is mountainous and except for a few valleys has bedrock at the surface or underlying a thin mantle of drift.

### REFERENCES

1. ARMSTRONG, J. E. (1954). Preliminary map, Vancouver North, British Columbia. Geol. Surv., Can., Paper 53–28.
2. ———— (1956). Surficial Geology of Vancouver Area, British Columbia. Geol. Surv., Can., Paper 55–40.

3. —— ((1956). Application of geology to soil problems in the lower mainland of British Columbia. N.R.C. Proceedings of Ninth Canadian Soil Mechanics Conference, Tech. Paper No. 41, pp. 11–19.

4. ——— (1957). Surficial geology of New Westminster map-area, British Columbia. Geol. Surv., Can., Paper 57–5.

5. ——— (1957). Geology of the Douglas-fir region of British Columbia. Forest Soils Committee of Douglas-Fir Region, Univ. of Wash., Seattle, pp. II–1–12.

6. — — (1960). Surficial geology of Sumas map-area, British Columbia. Geol. Surv., Can., Paper 59–9.

7. ——— (1960). Preliminary map, surficial deposits, Chilliwack map-area, British Columbia. Geol. Surv., Can., Prel. Map. 53–1959.

8. ———and Brown, W. L. (1953). Ground-water resources of Surrey Municipality, British Columbia. Geol. Surv., Can., Water Supply Paper No. 322.

9. ——— (1954). Late Wisconsin marine drift and associated sediments of the Lower Fraser valley, British Columbia, Canada. Bull. Geol. Soc. Amer., 65: 349–64.

10. Bostock, H. S. (1948). Physiography of the Canadian Cordillera, with special reference to the area north of the fifty-fifth parallel. Geol. Surv., Can., Mem. 247.

11. Fyles, J. G. (1956). Surficial geology of the Horne Lake and Parksville map-areas, British Columbia; Doctorate dissertation at University of Ohio.

12. ——— (1959). Surficial geology of Oyster River map-area, British Columbia. Geol. Surv., Can., Map 49–1959.

13.          (1960). Surficial geology of Courtenay map-area, British Columbia. Geol. Surv., Can., Map. 32–1960.

14. Halstead, E. C. (1958). Ground-water resources of Langley Municipality, British Columbia, Geol. Surv., Can., Water Supply Paper No. 327.

15. ——— (1959). Ground-water resources of Matsqui Municipality, British Columbia. Geol. Surv., Can., Water Supply Paper No. 328.

16. —— (1960). Ground-water resources of Sumas, Chilliwack, and Kent Municipalities, British Columbia. Geol. Surv., Can., Paper 60–29.

# GLACIAL DEPOSITS OF ALBERTA

## C. P. Gravenor and L. A. Bayrock

THE GLACIAL DEPOSITS of central and southern Alberta have been under study intermittently for some seventy-five years. Most of the early work was confined to the examination of sections along the major river courses and to broad reconnaissance studies. After the Second World War detailed mapping of the surficial deposits of Alberta was started by the Geological Survey of Canada and the Research Council of Alberta. To date approximately 37,000 square miles of central and southern Alberta have been mapped in detail. In addition, the Research Council of Alberta has mapped about 30,000 square miles of northern Alberta over the past two years by means of helicopter surveys. Thus, roughly one-quarter of the surface deposits of the Province have been mapped.

Soil maps covering parts of the Peace River country, central and southern Alberta and reconnaissance maps of northern Alberta have been prepared by the Alberta Soil Survey. Although in some cases the nomenclature used by soil scientists for the parent materials of the soils is not identical with that used by geologists, these maps are nevertheless extremely useful to the geologist and provide an excellent background for detailed geological study of surficial materials.

This paper is designed to give a generalized picture of the nature, distribution and age of the surficial deposits of Alberta. Undoubtedly many changes will be made as more detailed mapping is completed, but it is believed that this summary will serve as a basic framework to future studies.

### REGIONAL SETTING

*Preglacial Topography*

From the middle of the Tertiary period to the onset of glaciation during the Pleistocene epoch, Alberta was subjected to extensive erosion which resulted in the removal of hundreds of feet of Tertiary and Cretaceous sediments from the Plains region. The amount of material that has been removed can be estimated from the relief shown by gravel-capped erosion remnants which are found in many parts of the province. In southern Alberta (Figure 1) the Cypress Hills rise about 1,800 feet above the surrounding Plains surface and are capped by gravels of Oligocene age (Russell and Landes, 1940). In northern Alberta, the Swan Hills, Buffalo Head Hills, Clear Hills, Cameron Hills, Naylor Hills, Mount Watt, and the

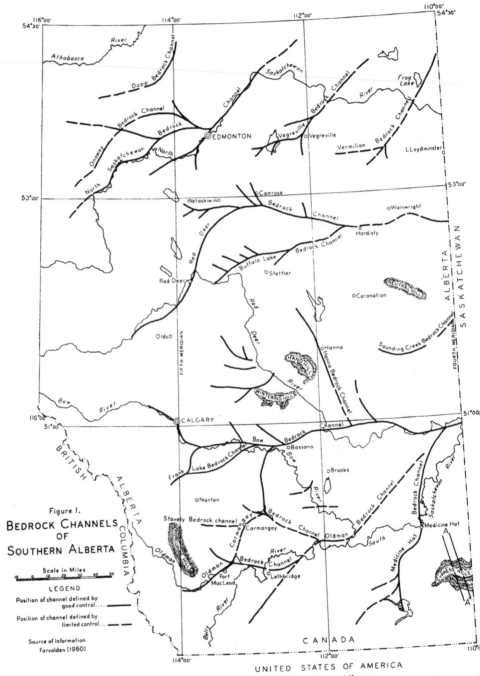

FIGURE 1. Bedrock channels of southern Alberta

Birch and Caribou Mountains all show local relief of over 1,000 feet and may form a part of the same erosion surface—the Cypress Plain—as that represented by the Cypress Hills in southern Alberta. A lower erosion surface which has a local relief of from 300 to 600 feet in southern Alberta is represented by the gravel-capped Hand and Wintering Hills and possibly by the Neutral Hills (Figure 1). Craig (1957) considers that the gravel cap on the Hand and Wintering Hills may be correlative with the Flaxville gravels and hence may be Pliocene in age (Alden, 1932).

In most cases these gravel-capped erosion remnants are surrounded by broad pediment-like surfaces that slope away from the actual remnant to the major drainage courses (Figure 2). In some instances, these pediment-like surfaces have remnants of gravel that was derived from the reworking

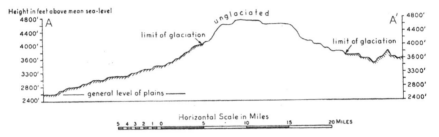

Figure 2.    Projected profile through Cypress Hills along line A-A' (Figure 1)

of the gravel caps on the higher surfaces. Judging from the shapes of the erosion remnants and of the surrounding surfaces, it is believed that the erosion took place under arid to semi-arid conditions and was a process of pediplanation rather than one of peneplanation.

The topography of the Plains immediately before glaciation was much the same as it is today, except for the thin veneer of glacial materials and modifications to the drainage systems caused by glaciation (Farvolden, 1960). Figure 1 shows the bedrock channels which were present before glaciation and which are now buried under drift. Although in some places present-day rivers follow the courses of the earlier channels, in most cases there is little relationship between the two sets of drainage patterns.

These channels are commonly floored with gravel and sand. They have been located by the use of water-well drilling logs, seismic "shot hole" logs obtained from oil companies, and seismic profiles and drilling carried out by the Research Council of Alberta. The gravels are composed primarily of quartzite with lesser amounts of argillite, limestone, basic volcanics, arkose, chert and local bedrock (e.g. Horberg, 1952, Rutherford, 1937). The apparent lack of granites and metamorphic rocks suggests that the gravels were deposited before glaciers moved in from the Canadian Shield.

In southwestern Alberta, and in the area west of Edmonton, the bedrock channels display an intersecting pattern which is possibly the result of drainage disruption by glacier blocking. The number of control points in southwestern Alberta is somewhat limited, however, and it is possible that the pattern may assume more "normal" characteristics when sufficient information becomes available to plot bedrock topography in this part of the province.

### Direction of Glacier Advance

In central and southern Alberta the gravels that floor the bedrock channels—Saskatchewan gravels and sands (Rutherford, 1937)—are overlain by two or more till sheets which are separated by boulder pavements and bedded silts and sand (e.g. Warren, 1954; Horberg, 1952). These till sheets were deposited from glaciers which moved into Alberta from the Northwest Territories (Keewatin centre) and from the Rocky Mountains (Cordilleran centre). The directions of ice movements associated with the earlier glaciations are not known, but there is a certain amount of information on the directions of advance and retreat of the last major glacier.

The regional directions of advance of the last ice are shown on Figure 3. As might be expected, much of the evidence pertaining to directions of advance has been obscured by local rejuvenations of the ice during its retreat. It is possible, however, to give a generalized picture of directions of advance from the following evidence: (1) drumlin and fluting directions, which were not disrupted during the ice retreat: the directions north of Rocky Mountain House and those that occur on the tops of erosion remnants (e.g. Birch Mountains, sixty miles northwest of McMurray) fall into this category; (2) distribution of erratics, such as oil sand from the McMurray area and quartzite erratics from the Mountains; (3) analysis of the altitudes of the highest glacial deposits in southwestern Alberta and around the Cypress Hills.

Evidence of this type is open to question, in that there is no reliable method of distinguishing the relative ages of drumlins and flutings, and because some of the erratics may have been shifted from their original positions by rejuvenation of the retreating ice. Nevertheless, there appears to be an overall consistency in the various lines of evidence; hence it is believed that the regional advance directions given in Figure 3 are reasonably correct.

Examination of the directions of advance show that the glacier which moved out from the Northwest Territories crossed northern and central Alberta in a southwesterly direction. In western Alberta, the Keewatin glacier met the ice which moved out of the Mountain passes and the two ice masses joined and flowed in a southeasterly direction roughly parallel to the Foothills. The suggested southeasterly ice-movement direction in southern Alberta is derived from the distribution of quartzite erratics (Stalker, 1956) and also from the altitudes of the highest glacial deposits from the last ice in southwestern Alberta and around the Cypress Hills (Figure 3). Northwest of Calgary, the highest deposits of Keewatin origin (from the last ice) lie at an altitude of about 4,200 feet (Tharin, personal communica-

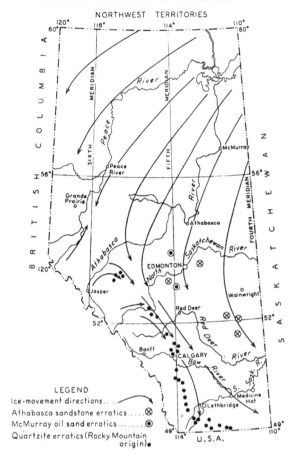

FIGURE 3.   Generalized ice-advance directions in Alberta

tion). In the Cypress Hills, the upper limit of glacial deposits is at an altitude of 4,100 feet on the north side of the hills and at an altitude of 3,700 feet on the south side. Thus, it is evident that the ice surface was almost flat across southern Alberta and hence the flow direction must have been towards the south.

### Glacier Retreat

The ice-movement directions shown on Figure 4 are a combination of both direction of advance and retreat. Many of the directions in southern and central Alberta show a fan-like distribution and this feature, in conjunction with an analysis of transverse elements, suggests that most of these directions were created by local rejuvenations of the ice during general retreat. In eastern and northern Alberta, the ice-movement directions can

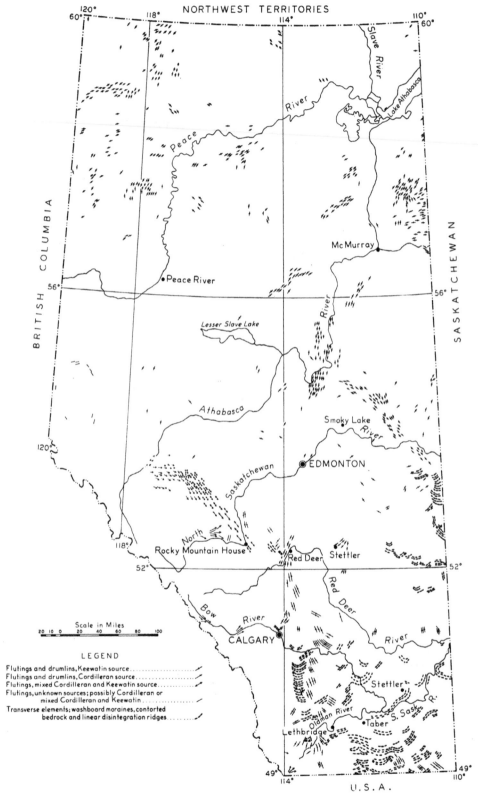

FIGURE 4. Ice movement directions in Alberta

be roughly divided into two groups—first, those which indicate a south-westerly movement and, second, those which indicate a southeasterly movement.

Many of the features showing southwesterly trends are found on the tops of highland erosion remnants; it is believed that they were formed during the advance of the last major glacier. The southeasterly-trending features, such as the Lloydminster trend, are composed of long narrow belts of fluting which possibly record late surges of ice in a southeasterly direction (Gravenor and Meneley, 1958) (Plate 1). The transverse elements shown on Figure 4 consist largely of washboard moraine (Plate 2), linear disintegration ridges (Plate 3), and contorted bedrock ridges (Gravenor et al., 1960). These elements were developed near the margins of retreating ice lobes and are all apparently related to thrusting in the ice margin (Elson, 1953).

One of the more significant points brought out by this study is that the last ice in Alberta retreated in a northerly and northeasterly direction. A similar conclusion on the directions of ice retreat is also indicated by the locations of ice-marginal meltwater channels which mark successive positions of the terminus of the retreating ice mass (e.g. Gravenor and Bayrock, 1956).

The nature of the retreat was apparently one of large-scale down-wasting leading to stagnation, followed by local rejuvenation which resulted in small lobes or fingers being sent out from the main ice mass. From analysis of the fan-shaped nature of the ice-movement directions in southern and central Alberta, in conjunction with the distribution of transverse elements, it can be suggested that the cycle of stagnation-rejuvenation must have occurred many times during general retreat. Certain of the rejuvenations were of considerable magnitude (Gravenor et al., 1960, p. 6) and it is possible that these may correlate with known readvances in the Great Lakes region during the retreat of the classical Wisconsin ice. Establishment of such correlations, however, will depend primarily upon the finding of suitable material for radiocarbon dating.

### SURFICIAL DEPOSITS

The surficial deposits of the province have been grouped under three main headings and their areal distributions are shown in Figure 5. Soil reports—particularly on areas in northern Alberta—have proven to be by far the largest single source of information and this fact indicates the value of such reports in compiling regional maps showing surficial materials.

### Ground Moraine and Hummocky Moraine

Parts of central and southern Alberta are characterized by broad areas of flat to gently rolling ground moraine. In some places, the ground moraine is almost completely featureless and at other places there is extensive

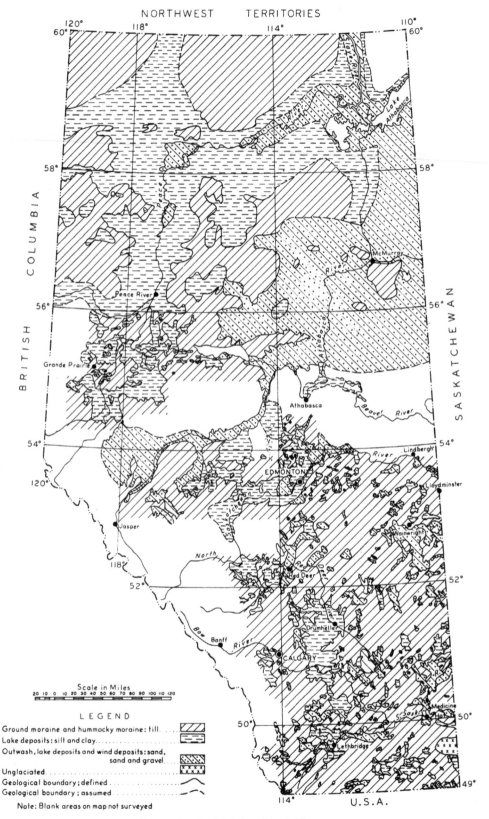

FIGURE 5. Surficial deposits of Alberta

development of till ridges and washboard moraines (Gravenor *et al.*, 1960). These features are commonly of low relief—less than ten feet—and their presence is best detected from air photographs (Plates 2, 3).

The ridges of eastern Alberta are composed mainly of till and may or may not have included pockets of stratified materials. Till ridges have been occasionally noted to grade into ridges composed almost entirely of stratified materials. The ridges vary in height from 3 to 35 feet, in width from 25 to 300 feet, and in length from a few yards to several miles. An important characteristic of the ridges is that, in general, two sets intersect at acute or right angles. Such intersection of ridges forms a "waffle," diamond or box pattern (Plate 3). In eastern Alberta the two sets of ridges are developed at right angles to and parallel to the direction of glacier motion, the former being more prominent. It is believed that the material now forming the ridges accumulated in crevasses in stagnant ice.

The ground moraine grades commonly into irregular to elongate tracts of hummocky moraine which are composed of knobs and kettles, linear elements in the form of till ridges or chains of knobs, and moraine plateaux (Plate 4). The knobs are composed largely of till but, like the till ridges, may have included lenses of stratified materials. Many of the knobs have a "doughnut" shape which is due to the presence of a circular undrained depression on the top of each mound. In some places the knobs are aligned to form beaded ridges. These beaded ridges are often gently curved in outline and, like other transverse elements, mark a position of the ice margin during retreat.

Moraine plateaux are relatively flat areas in the hummocky moraine (Plate 4). The plateaux are composed predominantly of till but in some places there is a thin cover—two to ten feet thick—of lacustrine silts and clays at the surface. The edges of the plateaux may be delineated by minor till ridges, referred to as "rim ridges," but in eastern Alberta such rim ridges are the exception rather than the rule. Although opinions differ on the mechanics of the formation of hummocky moraine it is generally agreed that the various features were developed during the down-wasting or stagnation phase of deglaciation.

## Composition of Till

Information on the mineralogical and mechanical composition of till in Alberta is somewhat limited, but there are sufficient data to give a general idea of its characteristics. In a recent study of the heavy mineral content of till in eastern Alberta, Bayrock (1960) found that the heavy minerals make up about 1.7 per cent by weight of the total till and that the bulk of the heavy minerals was derived from Canadian Shield rocks. Quantitative analysis of the heavy minerals in the bedrock from the same area of eastern Alberta, and also in rocks from the Precambrian Shield in northeastern Alberta, shows that the local bedrock contains only 0.1 per cent by weight of heavy minerals whereas the Precambrian rocks contain 8.7 per cent by weight of heavy minerals. Thus the till in eastern Alberta is composed of

PLATE 1. Flutings which record a southeasterly ice-movement direction in the Lloydminster area. Location: Twp. 57, R. 7, W. 4th meridian.

PLATE 2. Washboard moraine. Arrow indicates ice-movement direction as taken from flutings. Note washboard moraines are superimposed on flutings. Location: Twp. 15, R. 27, W. 4th meridian.

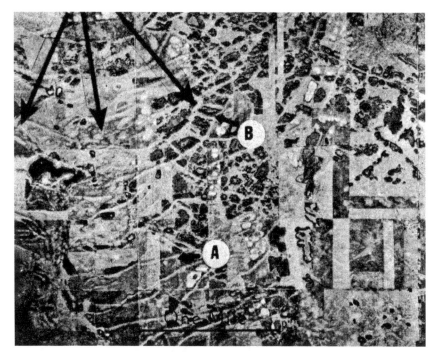

PLATE 3. Till ridges. Long arrows show direction of ice-movement. A: arcuate ridges developed at right angles to ice-movement direction. B: "box" or "Waffle" pattern formed by intersection of ridges. Location: Twp. 48, R. 1, W. 4th meridian.

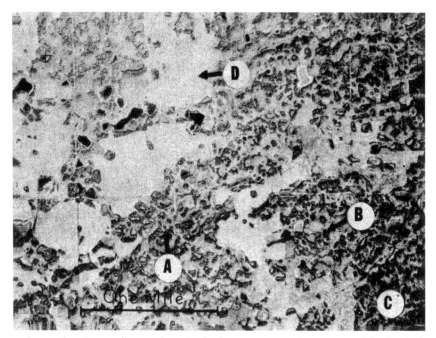

PLATE 4. Hummocky moraine. A: knobs; note vegetation at top of knob which marks the site of an enclosed depression. B: water-filled kettle. C: beaded edge; trend is in a northeasterly direction. D: moraine plateau. Location: Twp. 46, R. 12, W. 4th meridian.

about 80 per cent local bedrock. Statistical treatment of the results of the study indicates that the surface till of eastern Alberta cannot be classified by means of heavy mineral assemblages, even though large discrepancies in composition may exist.

Clay minerals studies made on the till of eastern Alberta indicate that in the fine fraction ($< 0.2 \mu$), montmorillonite is the major clay mineral and that lesser amounts of illite and kaolinite are present. In the coarse clay fraction ($2.0-0.2\mu$), illite is the dominant clay mineral with lesser amounts of montmorillonite, kaolinite and chlorite present. Studies made on the clay minerals from till in the area just east of Edmonton, in northwestern Alberta and in the McMurray area, indicate that montmorillonite is the major clay mineral, with illite subsidiary. Kaolinite is present in only minor amounts. Investigations of the composition of lake and alluvial materials show that they contain the same suite of clay minerals as is found in till but in varying proportions.

It is of interest to note that the mineral chlorite occurs in eastern Alberta, but has not been found in significant amounts in the tills of northern Alberta. In eastern Alberta the tills sampled are underlain by Bearpaw shales which are known to contain chlorite (Byrne and Farvolden, 1959). This, in addition to the results of the heavy mineral studies, suggests that the tills are largely local in origin. Samples of Cordilleran till from the Calgary area (Tharin, personal communication) and from the Rocky Mountain House area indicate that this till has a much lower montmorillonite to illite ratio than has adjacent till of Keewatin origin. In addition, the Mountain tills in general have a higher carbonate content than the Keewatin till and, as would be expected, there are significant differences in pebble contents.

Over the past few years, the Research Council of Alberta has been involved in a programme to show variations in the mechanical composition of the surface till. This project has been designed to show regional and local variations in till that will be of use in soil classification and in geological studies. The programme is far from complete but the preliminary results include some interesting facts which can be summarized as follows:

*Local variations*

Mechanical analyses of fifty till samples of the Wainwright area in eastern Alberta (Figure 6) show that locally the tills have a remarkably uniform texture. Roughly 75 per cent of the samples can be classified as sandy clay loam and the bulk of the remainder as loam or clay loam. The mean composition of the fifty samples is as follows (classification based on the soil texture triangle used by soil scientists in Canada and the United States): sand, 47 per cent; silt, 25 per cent; and clay, 28 per cent.

*Regional variations*

Regional variations in till, plotted in terms of clay content, are shown in Figure 6. This map shows that in general the tills of southeastern Alberta are less clayey (and hence more sandy and silty) than those of eastern

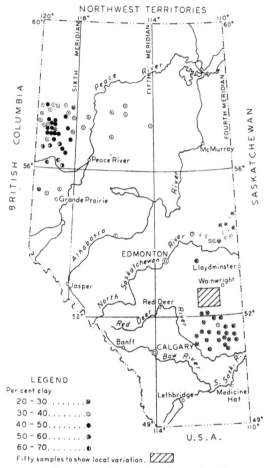

FIGURE 6. Textural variation in till

Alberta, which in turn are less clayey than those of north-central and northwestern Alberta. These variations are a function of the nature of the local bedrock (which is more clayey in the Peace River country), the direction of ice advance and the nature of pre-existing glacial deposits.

*Lake deposits*

The most striking feature of the surface materials of Alberta is the presence of vast areas of glacial lake deposits in the northern part of the province. This large accumulation of lake deposits in northern Alberta, and the more sporadic occurrence of lake deposits in central and southern Alberta, is a consequence of the direction of ice retreat and of regional topography. In southern and central Alberta, the regional slope is towards the east; hence lakes that were ponded between the retreating ice margin and the higher land to the west were quickly drained through outlets around the terminus of the northward-retreating ice. Thus there was not the opportunity for the entrapment of large quantities of water in southern Alberta; hence the ice-marginal lakes were relatively small and short-lived.

North of the Edmonton area the regional slope changes from east to northeast, and the major present-day rivers such as the Athabasca and Peace flow to the Mackenzie River. During retreat of the ice from this region, the meltwaters from both the retreating Mountain ice and the retreating Keewatin ice were ponded between the Keewatin ice terminus and higher land to the southwest. As the ice continued to retreat, successively lower outlets were found around the southern and eastern limits of the retreating ice. One of the first of the outlet routes drained the lake waters to the south into the North Saskatchewan River near the town of Lindbergh (Figure 5). Upon further retreat the North Saskatchewan River outlet was abandoned in favour of a lower outlet along the course of the present Beaver River (Figure 5). As lower ground exists north of the Beaver River, undoubtedly other outlets will be found when this part of the province is mapped.

As lower outlets to the east were uncovered, the lake-levels dropped. Hence the lake deposits shown in Figure 5 represent a composite picture of the lakes which existed during the ice retreat. When more is known about the outlets and their elevations, it will be possible to outline the lakes associated with each outlet and in this way to gain a better understanding of the local glacial history.

The lake deposits in central and northwestern Alberta are composed largely of silt and clay, and stratification varies from excellent varving to rude bedding. Ice-rafted boulders are commonly found in the lake deposits of central and northwestern Alberta. In northeastern Alberta and parts of central and southern Alberta, the lake deposits consist largely of sand. The sand originated in part from the meltwaters and surface run-off from the retreating Keewatin ice and in part from glacial meltwaters and surface run-off from the Mountains (Hughes, 1958). A large part of this sand has been modified by wind action which resulted in the formation of U-shaped dunes (Plate 5). Except for small areas of active sand in northeastern Alberta, the dunes are now fixed by vegetation. Odynsky (1958) has made a study of the major dune areas of the province and has plotted the wind directions responsible for the dune formation. It appears that the dunes originated under similar wind conditions to those which exist at present.

### AGE AND CORRELATION OF KEEWATIN DRIFT SHEETS

The Pleistocene succession in central and southern Alberta has been under study for many years. As a result of these studies, many sections have been described and theories formulated to fit the known facts. Although it is generally recognized that Alberta has been glaciated more than once, no buried soils or other evidence of any long period between glacial deposits have been found; consequently there has been little agreement among the various participating geologists on the age and correlation of the glacial deposits.

In recent years, the history of the advance and retreat of the Wisconsin ice in the Great Lakes area has been documented by hundreds of radiocarbon dates. It was hoped that a similar achievement would be possible in western Canada, but results to date have been disappointing. Only four effective dates have been obtained for Alberta, three of which are beyond the limit of the dating equipment used and the fourth of which was made on a piece of wood in outwash and is probably of minimal significance. Nevertheless, the dates are worth recording as they do have some significance in unravelling the glacial history of western Canada.

The first of these dates was made on a spruce log found in the surface drift in the Smoky Lake district, northeast of Edmonton. This wood was originally dated at 21,600 ± 900 years B.P. (Gravenor and Ellwood, 1956). Later, a sample of the same wood was submitted to the Research Council of Saskatchewan (sample S92) where it was dated at > 31,000 years (K. J. McCallum, personal communication). This later date is the one now accepted for this material. The second and third dates were obtained from wood samples found in intertill sands in southern Alberta, along the Oldman River. These samples have provided ages of > 30,000 years and > 26,000 years (Stalker, 1958; Broecker et al., 1956).

The fourth date of 11,000 years was obtained from a piece of willow wood found forty feet below the surface in crossbedded outwash sands at Taber, in southern Alberta. From regional evidence it is reasonably certain that the ice front was not far away during the deposition of these sands; hence it is believed that this date marks the time of retreat of the last ice from southern Alberta. This dating was also made by K. J. McCallum at the Research Council of Saskatchewan and carries their number S68.

Although the evidence is meagre, the dates from Smoky Lake and Taber suggest that the last major glacier advanced across central Alberta more than 31,000 years ago, and that central and southern Alberta remained under ice cover until about 11,000 years ago. This advance and retreat probably coincides with the advance and retreat of the last major ice advance in eastern Canada which began about 48,000 years B.P., and where, aside from a retreat at about 28,000 years, the ice remained until Recent time (Dreimanis, 1959).

At the base of the Keewatin drift in Alberta, the youngest known deposits —other than Cordilleran drift of unknown age—are the Saskatchewan gravel and sands which were deposited along the bedrock channels mentioned previously in this paper. These gravels are generally called "preglacial," in the sense that they predate the overlying drift sheets and are assumed to be of late Tertiary or very early Pleistocene age. Dating of the gravels can be attempted by two techniques; first, by consideration of the time of their development relative to ancient erosion surfaces; and second, from fossils found within the gravels.

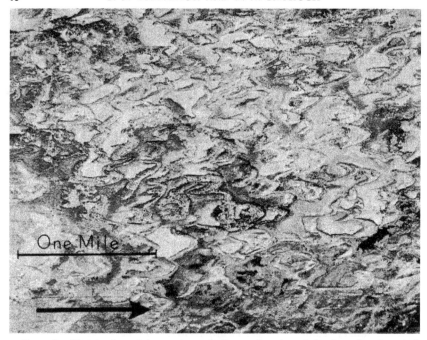

PLATE 5. U-shaped sand dunes. Arrow indicates direction of the dune-forming wind. Location: Twp. 70, R. 4, W. 6th meridian.

In a study of the physiography of Montana and southern Alberta, Alden (1932) noted four distinct erosion levels which he named, from oldest to youngest (highest to lowest), Cypress, Flaxville or No. 1 bench, and No. 2 and No. 3 benches, respectively. The Cypress Plain is generally assigned to the Oligocene, the Flaxville to the Pliocene, the No. 2 bench to the early Pleistocene and the No. 3 bench to the late Pleistocene. Drift of Cordilleran origin has been found capping the Flaxville surface in areas of northern Montana and southern Alberta (Alden, 1932; Horberg, 1954). This drift— the Kennedy drift  has a very deep soil profile and is probably of early Pleistocene age, possibly Kansan (Horberg, 1954). Inasmuch as there has been considerable erosion of the Plains since the deposition of the early Pleistocene Mountain drift, Horberg (1954) suggested that the No. 2 and No. 3 benches in the Waterton area of southwestern Alberta were eroded during the interval between the Kansan and Wisconsin glaciations.

Rutherford (1937) has reported the finding of mammoth remains (tusks and teeth) from several localities in Alberta and the Northwest Territories. Although the stratigraphic positions of the finds were not fully documented, he believed they were derived from Saskatchewan gravel and sand. Small collections of invertebrate faunas—predominantly gastropods -have been obtained from the Saskatchewan gravel and sands in the Edmonton area and are currently under study. This work is not sufficiently advanced to permit definite age determinations to be made, but it is believed that the fauna are of Pleistocene age and hence do not conflict with ages inferred from other methods of study.

Thus, available evidence provided from an analysis of erosion surfaces and from faunal data indicates that it is by no means certain that the gravels found below the drift are "preglacial" in the sense that they are late Tertiary or very early Pleistocene in age. From the present state of knowledge, it is not possible to assign a definite age to the gravels, but undoubtedly these gravels can be dated from fossil evidence; it is only a matter of time before adequate material is found and identified. The dating of these gravels is of obvious significance in determining the age of the drift sheets in Alberta.

### ACKNOWLEDGEMENTS

In compiling the information contained within this report the writers have had to rely heavily on the support of co-workers in the Research Council. To these geologists, draftsmen and soil scientists they express their thanks. They are especially indebted to S. Pawluk of the University of Alberta for offering unpublished information on the mineralogy and mechanical analysis of tills, to J. C. Tharin for unpublished information on mineralogy and geology of the Calgary area, to R. N. Farvolden for unpublished information on the bedrock channels, to the Geological Survey of Canada for providing unpublished data on the Red Deer area, and to S. Mathur for aid in compiling the map of surface deposits.

### REFERENCES

1. ALDEN, W. C. (1932). Physiography and glacial geology of eastern Montana and adjacent areas. U.S. Geol. Surv., Prof. Paper 174, 133 pp.
2. BAYROCK, L. A. (1960). Heavy minerals in till of central Alberta. Res. Coun. Alberta, unpublished manuscript.
3. BRETZ, J. H. (1943). Keewatin end moraines in Alberta, Canada. Geol. Soc. Amer. Bull., 60: 303–30.
4. BROECKER, W. S., KULP, J. L. and TUCEK, C. S. (1956). Lamont natural radiocarbon measurements III; Science, 124, no. 3213: 157.
5. BYRNE, P. J. S. and FARVOLDEN, R. N. (1959). The clay mineralogy and chemistry of the Bearpaw formation of southern Alberta. Res. Coun. Alberta Bull. 4: 44 pp.
6. CRAIG, B. G. (1957). Surficial geology, Drumheller (East Half) Alberta. Geol. Surv. Canada, Map 13–1957.
7. DREIMANIS, A. (1959). Proposed local stratigraphy of the Wisconsin glacial stage in the area south of London, southwestern Ontario. Contrib. Dept. of Geology, Univ. of Western Ontario, no. 25: 24–30.
8. ELSON, J. A. (1953). Periodicity of deglaciation in North America, Part II, Late Wisconsin recession. Geografiska Annaler, 35: 95–104.
9. FARVOLDEN, R. N. (1960). Bedrock channels in southern Alberta. Res. Coun. Alberta Bull. (in preparation).
10. GRAVENOR, C. P. and BAYROCK, L. A. (1955). Use of indicators in the determination of ice-movement directions in Alberta. Geol. Soc. Amer. Bull., 66: 1325–8.

11. ——— and            (1956). Stream-trench systems in east-central Alberta. Res. Coun.
    Alberta, Prelim. Rept. 56–4, 11 pp.

12.            and ELLWOOD, R. B. (1956). A radiocarbon date from Smoky Lake, Alberta.
    Res. Coun. Alberta, Prelim. Rept. 56–3, 17 pp.

13. ——— and MENELEY, W. A. (1958). Glacial flutings in central and northern
    Alberta. Amer. J. Sci., 256: 715–28.

14. ——— GREEN, R. and GODFREY, J. D. (1960). Air photographs of Alberta. Res.
    Coun. Alberta Bull. 5, 38 pp.

15. HORBERG, LELAND (1952). Pleistocene drift sheets in the Lethbridge region, Alberta,
    Canada. J. Geol., 60: 303–30.

16. ——— (1954). Rocky Mountain and continental Pleistocene deposits in the Water-
    ton region, Alberta, Canada. Geol. Soc. Amer. Bull., 65: 1093–150.

17. HUGHES, G. M. (1958). A study of Pleistocene Lake Edmonton and associated
    deposits. Univ. of Alberta, M.Sc. thesis.

18. JOHNSON, W. A. and WICKENDEN, R. T. D. (1931). Moraines and glacial lakes
    in southern Saskatchewan and southern Alberta, Canada. Roy. Soc. Canada Trans.,
    25, Sec. 4: 29–44.

19. ODYNSKY, WM. (1958). U-shaped dunes and effective wind directions in Alberta.
    Can. J. Soil Sci., 38: 56–62.

20. RUSSELL, L. S., and LANDES, R. W. (1940). Geology of the southern Alberta plains.
    Geol. Surv. Canada, Mem. 221, 223 pp.

21. RUTHERFORD, R. L. (1937). Saskatchewan gravels and sands in central Alberta. Roy.
    Soc. Canada Trans., 31, Sec. 4: 81–95.

22. STALKER, A. MacS. (1956). The erratics train foothills of Alberta. Geol. Surv.
    Canada, Bull. 37, 28 pp.

23. ——— (1958). The Kipp section, significant new information. J. Alberta Soc. Pet.
    Geol., 6: 252.

24. WARREN, P. S. (1954). Some glacial features of central Alberta. Roy. Soc. Canada
    Trans., 48, Sec. 4: 75–85.

Among the sources of information used in the preparations of the maps appearing with
this paper were reports of the Alberta Soil Survey and the Research Council of Alberta,
and maps from the Geological Survey of Canada. Also used were air photo mosaics by
the Department of Lands and Forests, Alberta.

# SOILS OF THE LAKE AGASSIZ REGION*

## John A. Elson

GLACIAL LAKE AGASSIZ is represented by lacustrine deposits distributed over an area estimated very roughly at 180,000 square miles (Figure 1). Several glacial lakes in connecting basins are included. The lake never covered this vast area at one time but expanded to the north and northeast as it shrank in the south.

Space limitations preclude a history of the exploration of the lake here, but a bibliographical outline is included at the start of the References for the interested reader. Specialized studies are referred to in the body of this paper. Several references not mentioned in the text are listed in the bibliography for the sake of completeness.

Glacial Lake Agassiz was first thought to have formed when the last continental ice sheet retreated northward and formed a closed basin that filled with meltwater and overflowed southward (Upham, 1895). It was later demonstrated by Tyrrell (1896) and Johnston (1916) that a lake formed in this basin when ice advanced. Elson (1957a) postulated an early lake (Lake Agassiz I), formed in the manner advocated by Upham, which was followed by an interval of complete or nearly complete drainage, through eastern outlets, and subaerial erosion. A glacier advance (Valders) or crustal uplift closed the eastern outlets and formed Lake Agassiz II. This lake first discharged southward through the Minnesota and Mississippi river systems, then, when the ice margin again retreated, expanded northeastward and discharged through eastern and then northern outlets. Radiocarbon datings indicate that the dry interval between Lake Agassiz I and II occurred about 10,000 years ago and that Lake Agassiz II discharged southward from about 9,000 years ago or earlier to 8,000 years ago. It was drained perhaps 7,000 years ago (Rubin and Alexander, 1960).

Glacial Lake Agassiz is known mainly from widely scattered records and its history is reconstructed from many tenuous interpolations. At the time of writing glacial events in Manitoba are not reconciled with those in Minnesota described by Wright (1956) and in North Dakota and Minnesota by Lemke and Colton (1958).

### EXTENT AND GEOLOGICAL SETTING OF LAKE AGASSIZ

The northern and eastern boundaries of Lake Agassiz (Figure 1) are subject to revision whenever a new geological report of those areas is

*Published by permission of the Director, Geological Survey of Canada.

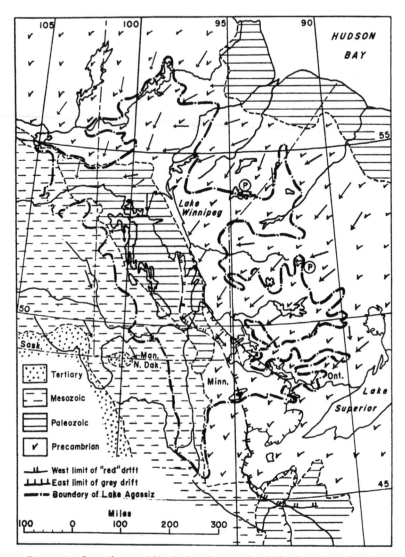

FIGURE 1. Part of central North America showing bedrock geology, directions of the last glacial movements, and the boundary of Glacial Lake Agassiz. Circled letter "P" indicates occurrence of Palaeozoic erratics from the Hudson Bay lowland. Arrows show direction of ice flow indicated by striations and streamlined features.

published. The existence of the lake is inferred mainly from reports of varved clays or similar sediments. In much of this region the lake margin consisted of an ice front part of the time, or had the intricate pattern of series of archipelagos. Wave erosion must have been greatly impeded in this island-filled part of the lake with its substantial seasonal fluctuation of water level. As a result, shore features are rare and usually occur only on high prominences of glacial drift.

In western and southern Manitoba, North Dakota, and Minnesota, shore features are well developed and the lake margins clearly defined. The tilting of the strandlines could be studied by Johnston (1946) because some beach ridges can be traced for literally hundreds of miles. The southern outlet was through the valley now occupied by Lake Traverse. Eastern outlets suggested by Antevs (1931) and Elson (1957a), and a northern outlet suggested from airphoto interpretation by Elson, have yet to be confirmed by field studies.

The soils of Lake Agassiz (Table I) are dominantly silt and clay overlying glacial till. The character of the till depends on the underlying bedrock and the direction of glacial transport. The lake sediments are derived from more remote glacier margins and differ from the till in composition.

| Soil type | Age | Nature | Distribution |
|---|---|---|---|
| Post-glacial and Recent deposits | | | |
| Organic soil | Post-glacial and Recent | Peat and muck in local basins; 3 to 30 feet thick | Mainly in the Precambrian area |
| Alluvium | Post-glacial to Recent | (1) Sand and gravel | River terraces |
| | | (2) Silt with sand, gravel and clay | Modern flood plains |
| | | (3) Gravel and sand, some clay, silt in natural levees, clay in back swamps | Alluvial fans near the foot of the Manitoba Escarpment |
| Aolian sand | Late glacial to Recent | Well sorted medium grained sand | Mainly on delta and shore deposits of Lake Agassiz |
| Lake deposits | | | |
| Beaches | Lake Agassiz | Well sorted sand and gravel | Abundant along west and south sides of Lake Agassiz; Whitemouth area of southeastern Manitoba |
| | | Medium grained sand | Inner part of the lake basin, especially on the west side |
| Near shore (littoral) deposits | Lake Agassiz | Silt and fine sand | Near the western margin of the lake basin; Whitemouth area; Lac Seul, Trout Lake area |

| Soil type | Age | Nature | Distribution |
|-----------|-----|--------|--------------|
| Lake deposits (cont'd) | | | |
| Deltaic deposits | Lake Agassiz I | Gravel, sand and silt | Southern part of west side of the lake basin |
| Deep water deposits | Lake Agassiz II | Laminated (varved) silt and clay | Western Ontario and northern Manitoba; in several extensive belts and numerous small basins |
| | Lake Agassiz II | Poorly laminated to massive clayey silt and clay; sandy at base, fossiliferous | Red River valley |
| | Lake Agassiz I | Thinly laminated to massive clay | Red River valley |
| Lag concentrate | Lake Agassiz I and II | Scattered to abundant residual boulders and gravel | Between Lake Winnipeg and the Manitoba Escarpment; Wave-cut terraces throughout the basin |
| Problematic gravel deposits | | | |
| Outwash, alluvium, or sub-aqueous moraine (?) | Post-glacial, pre-Agassiz (?) or early Lake Agassiz | Thin, lenticular, elongate bodies of gravel and sand | Red River valley; known only from borings; underlie clay of Lake Agassiz I |
| Glacial deposits | | | |
| End moraines | Valders and younger | Sandy till of Precambrian provenance; gravel and sand | Western Ontario and northern Manitoba |
| | Late Mankato | Mainly calcareous sandy to clayey till | Southern Manitoba, North Dakota and Minnesota |
| Ice-contact stratified drift | Mainly Valders some late Mankato | Sand and gravel of variable sorting in eskers and kames | Mainly western Ontario |
| Ground moraine | Valders | Sandy till of Precambrian provenance | Western Ontario and northern Manitoba |
| | Late Mankato | Grey, calcareous sandy to clayey till | Southern Manitoba, North Dakota and western Minnesota |
| | Mankato and earlier | Buff to red non-calcareous sandy till | Southeastern Manitoba, southwestern Ontario, and northern Minnesota |
| Tills and interglacial deposits | Mid-Wisconsin and older | Various | Known from wells in the Red River valley |

Precambrian rocks underlie the northern and eastern part of Lake Agassiz (Figure 1). Palaeozoic rocks, mainly limestone and dolomite, and Mesozoic rocks, chiefly Cretaceous shales, underlie the west-central, and southern parts respectively. At the unconformity below the Cretaceous rocks, the deeply weathered Precambrian rocks are covered by residual clays. Archaean rocks beneath the Palaeozoic sediments were weathered sufficiently to form iron ore from siliceous iron formation. Locally these weathered zones were incorporated into the glacial drift.

Bedrock debris was distributed in two directions by glacier flow (Figure 1). Southwestward movement distributed Precambrian debris (Patrician "red" drift) over the eastern half of the basin. A later southeastern flow transported Palaeozoic and Mesozoic rock debris (Keewatin grey drift) 170 to 220 miles across the red drift, in the southern part of the basin. The changes in direction of ice movement may have resulted from shifting of centres of glacial outflow and did not necessarily require deglaciation and the reconstitution of ice-sheets.

Ice flows inferred from striations and streamlined features in the northern part of the basin occurred later than those recorded by drift lithologies in the south. Lithologies of the northern drift are less distinctive partly because ice movements in any given direction were of relatively short duration, and consequently less debris was moved than in the south. Camsell (1904) and Wright (1927) reported Palaeozoic rocks transported about 160 miles southwest from the Hudson Bay lowland (Figure 1), but evidently they were not an important proportion of the drift. Satterly (1938a) found no Palaeozoic stones at Echoing Lake only 60 miles from Palaeozoic outcrops. The paucity of calcareous debris in this region may be due to its low resistance to abrasion and crushing in a matrix of much stronger Precambrian rock debris. In contrast, erratics of resistant felsite or greywacke from Cape Jones at the eastern entrance to James Bay were reported by Bell (1891) and he and Wright (1927) reported iron formation from there, or the Belcher Islands, about 600 miles from their known outcrops. About half this distance is occupied by Palaeozoic carbonate rocks.

## GLACIAL DEPOSITS

The ice movements transporting red and grey drift across the southern part of the Lake Agassiz basin were repeated several times during the Pleistocene epoch (Table II). Interglacial deposits within Lake Agassiz were reported by Johnston (1921, 1935), Rosendahl (1948, p. 291–6) and Allison (1932). These consist of gravel between overlying grey and underlying red drift, an older weathered grey till, organic remains—representing a climate like the present—below grey till, and an older weathered till, respectively. These occurrences do not necessarily represent the same interglacial interval. The organic matter described by Rosendahl may represent the St. Pierre Interval defined in the St. Lawrence lowlands by Terasmae (1958). Some of the weathered till may be older. Of prime concern here are the last movements of ice that occurred in Wisconsin time (Wright, 1956) and deposited the Patrician "red" drift which is overlain by Keewatin grey drift in northern Minnesota, southern Manitoba and the adjacent parts of Ontario.

TABLE II

GLACIAL DRIFTS OF THE LAKE AGASSIZ BASIN AND ADJACENT AREAS

| Age | Type and distribution |
| --- | --- |
| Wisconsin | |
| Valders | Precambrian-derived drift in western Ontario and northern Manitoba. |
| Late Mankato | "Young grey" drift covering south and southwestern parts of the lake basin, extending southeast of Lake of the Woods and south of Rainy Lake. |
| Mankato | "Young red" drift, in southwestern Ontario and northern Minnesota, extending into southeastern Manitoba. |
| Mid-Wisconsin | Grey drift in the south and west parts of the basin. |
| Interstadial (St. Pierre?) | Bronson, Minnesota, interstadial deposits. |
| Unnamed substage | Grey drift; may include Iowan of Leverett (1932); not identified, but should occur in Red River valley unless entirely removed by erosion. |
| Illinoian | "Old red" drift; generally very thin; of northern provenance; not identified in the Lake Agassiz basin. The southwestern limit of the Illinoian glaciation is uncertain but it probably extended over southern Manitoba and eastern North Dakota where a grey drift would have been deposited (Lemke and Colton, 1958). |
| Kansan | "Old grey" drift; calcareous blue-grey indurated and jointed till found in western Minnesota; a "red" type of drift in eastern Minnesota; not identified in the Lake Agassiz basin. |
| Nebraskan | Patches of deeply weathered drift containing much gravel and organic matter, the latter giving rise to the term "black till." Contains granite and limestone stones that may be locally derived. Occurs in southeastern Minnesota; not identified in the Lake Agassiz basin. |

*Patrician "red" drift*

The "red" drift is generally buff in colour, non-calcareous, sandy, and is derived from Precambrian rocks. Where the ice sheet crossed the iron ranges of Minnesota the drift is truly red. In northern Minnesota the red drift passes under grey drift at about the longitude of the east end of Rainy Lake (Figure 1). Its western limit is uncertain but may be slightly west of the Lake of the Woods. Johnston (1935, p. 17) suggests that it may continue a few miles west across Palaeozoic rocks and the pattern of glacial striations (Wilson *et al.*, 1958) supports this view.

The drift in western Ontario and northern Manitoba that was derived from Precambrian rocks can be considered red drift. The carbonate content ranges from 0.0 to 8.6 per cent, according to Kruger (1957) and Ehrlich and Rice (1955), and averages less than 2 per cent. Most of this is probably from the weathering of calcic feldspars and mafic minerals in basic volcanic rocks and greenstones (Table III). At Telford, just west of Lake of the Woods, Ehrlich and Rice found the rock particles in Patrician till to be 3 to 6 per cent limestone, 0.5 to 1.4 per cent shale and 93 to 97 per cent of other types, presumably mainly Precambrian granitoid rocks. Their mineralogical analyses are presented in Table IV. The high carbonate content at Telford is to be expected at the border of Palaeozoic rocks.

TABLE III

CHEMICAL ANALYSES OF TYPICAL ROCKS THAT HAVE BEEN INCORPORATED
INTO THE SURFACE DEPOSITS OF THE LAKE AGASSIZ REGION

| | Granite gneiss Fresh | Granite gneiss Weath. | Diabase Fresh | Diabase Weath. | Pierre shale | Niob. fm. | Altered basalt | Green-stone |
|---|---|---|---|---|---|---|---|---|
| | 1 | 2 | 3 | 4 | 5 | 6 | 7 | 8 |
| Silica | 63.61 | 60.61 | 57.65 | 43.35 | 79.94 | 20.67 | 45.44 | 50.65 |
| Alumina | 16.71 | 18.12 | 14.97 | 20.42 | 7.33 | 8.70 | 20.33 | 14.57 |
| Ferric oxide | 5.69 } | } 7.51 | 6.05 | 13.57 | } 2.28 | } 3.14 | 3.07 | 2.31 |
| Ferrous oxide | 2.78 } | | 4.23 | — | | | 5.40 | 11.59 |
| Magnesia | 1.63 | 1.14 | 4.00 | 5.17 | 1.08 | 0.50 | 4.76 | 4.34 |
| Lime | 4.03 | .03 | 4.30 | .27 | 1.07 | 33.23 | 12.12 | 6.70 |
| Soda | 1.68 | .54 | 2.01 | .35 | } 1.19 | } 1.20 | 2.33 | 2.45 |
| Potash | 2.49 | 3.56 | 3.18 | 5.46 | | | .45 | .59 |
| Moisture | — | .28 | .33 | 4.24 | } 7.00 | — | } 3.23 | } 2.67 |
| Combined water | .61 | 7.20 | 1.91 | 5.54 | | | | |
| Other | — | 1.30 | 2.24 | .92 | .11 | .49 | 2.76 | 4.42 |
| Ignition loss | — | — | — | — | — | 28.77 | — | — |
| | 99.23 | 100.29 | 100.87 | 99.29 | 100.00 | — | 99.89 | 100.29 |

EXPLANATION AND SOURCES:
1–4. Grout (1919).
5. Average of three analyses by Ries and Keele (1912). Moisture and combined water by
   difference.
6. Niobrara formation from Leary's, Manitoba, Ries and Keele (1912).
7 and 8. Uchi Lake area, western Ontario, Bateman (1940).

TABLE IV

APPROXIMATE MINERAL COMPOSITION OF RED AND GREY DRIFTS
IN SOUTHEASTERN MANITOBA
(after Ehrlich and Rice, 1955)

| | Composition in percentages | |
|---|---|---|
| | Red drift | Grey drift |
| Fine sand (.25–.10 mm.) in till | 20 | 13 |
| Light mineral fraction of fine sand | >96 | >96 |
| Composition of light fraction: | | |
| Quartz | 25 | 5–10 |
| Feldspars | 60 | 45 |
| Calcite and dolomite | 5–8 | 40 |
| Others | 5–15 | 3–8 |
| Heavy mineral fraction of fine sand | <4 | 4 |
| Composition of heavy mineral fraction: | | |
| Amphiboles | 50 | 55–65 |
| Pyroxenes | 10 | 5–20 |
| Garnet | 25 | 5 |
| Apatite | <2 | <2 |
| Iron oxides | 5 | 1 |
| Others | 5 | 15–20 |
| Total clay-size fraction (.002 mm.) of the till | <4 | 9 |
| Composition of the clay-size fraction: | | |
| Quartz | 10 | 5 |
| Feldspars | 10 | 5 |
| Montmorillonoids | 25 | 10 |
| Vermiculite | 5 | 5 |
| Chlorites | 10 | 10 |
| Micas | 10 | 10 |
| Illite | 10 | 5 |
| Mixed layer minerals | 10 | 10 |
| Kaolinite | 10 | 5 |

There is a local enrichment of iron oxides at Falcon, of up to 35 per cent of the heavy mineral fraction studied, that may be from local concentrations of iron in the Winnipeg sandstone which is the basal Palaeozoic rock there. At Black Island, 100 miles north-northwest of Falcon, the Winnipeg sandstone rests on a small body of Precambrian iron ore (Tyrrell and Dowling, 1901) and is cemented in places by secondary iron oxides. Clay minerals may be most common in the southern region of Patrician drift where deeply weathered Archaean rocks (Grout, 1919) were incorporated into glacial deposits. The chemical composition of this drift is discussed later.

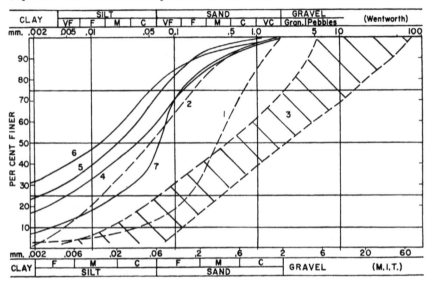

FIGURE 2. Cumulative grain-size distribution curves of tills in the Lake Agassiz basin. Dashed lines are "red" till, solid lines are grey till. Nos. 1 and 2: Falcon and Telford (Ehrlich and Rice, 1955); No. 3, envelope of Steep Rock Lake tills (Legget and Bartley, 1953); No. 4: Komarno (Ehrlich and Rice, 1955); Nos. 5 and 6: Rainy Lake area, analyses 16 and 18 (Johnston, 1915); No. 7: Upper Whitemouth area (Johnston, 1921).

The grain-size distribution in Patrician drift was investigated at Telford (Ehrlich and Rice, 1955) and at Steep Rock Lake (Legget and Bartley, 1953). It is generally a loose-textured, sandy, stoney till (Figure 2). The median grain size ranges from .05 to 2.5 mm, generally well within the range of sand. The effective size (the maximum particle size of the smallest ten per cent) ranges from .005 to .12 mm (very fine silt to very fine sand) indicating moderate permeability to water. The till is generally well graded (poorly sorted), and is locally suitable for road surfacing.

The western boundary of the Patrician red drift in Figure 1 is projected northward along the western limit of southwest-trending striations. In the area north and west of the latitude of the middle of Lake Winnipeg, it is as yet impossible to distinguish Patrician from Keewatin drift, because both ice sheets passed across similar types of bedrock. The heavy mineral content may prove to be diagnostic.

### Keewatin "grey" drift

In most exposures the Keewatin "grey" drift is a buff colour but where unoxidized it is grey. In northern Minnesota and adjacent Manitoba and Ontario, the grey color is due to a high proportion of calcareous material derived from Palaeozoic carbonate rocks to the west and north. In central Minnesota and North Dakota, it is mainly from Cretaceous shales. The texture is generally finer than that of Patrician till, ranging from sandy to clayey. Limestone and dolomite pebbles are dominant but Precambrian stones are present. The grey drift covers the Palaeozoic and Cretaceous rocks of the Lake Agassiz basin and extends eastward across the Precambrian area, overlying Patrician red drift in Minnesota and Ontario (Figure 1).

Where it overlies limestone and dolomite the till contains as much as 65 per cent carbonate (Ehrlich and Rice, 1955, Komarno). From about 70 to 120 miles from the area of limestone outcrop the carbonate content is 25 to 30 per cent, and near the periphery of grey drift in Minnesota, almost 300 miles from the bedrock source, it is only 10 to 20 per cent. The reduction in carbonate here depends mainly on the mixing in of new rock types and this in turn depends on the susceptibility of those rocks to glacial erosion; hence the reduction of the quantity of a rock type in drift is not entirely a function of the distance from the source.

The composition of the rock particles in till in the Palaeozoic outcrop area (Komarno) is about 75 per cent limestone and 25 per cent other rocks. Farther south the proportion of shale increases and that of limestone decreases. Mineralogical analyses by Ehrlich and Rice (1955) are presented in Table III. The clay minerals are probably more abundant farther south where they derive from Cretaceous shales and the weathered Precambrian "granite." The chemical composition of grey drift is discussed later.

Grain-size data for the grey drift (Figure 2) were published by Johnston (1915) and Ehrlich and Rice (1955). It usually contains more silt and clay than Patrican drift; median grain sizes range from .01 to .07 mm. The effective grain size is generally well within the clay-size range, which indicates impermeability to water. The grey till is commonly well graded.

Grey drift underlies the lake deposits in the Red River basin and ranges from a few feet thick at the latitude of Winnipeg to about 200 feet at places in North Dakota and Minnesota.

### Red and grey drifts contrasted

In summary, the Patrician red drift is a sandy non-calcareous drift, and the Keewatin grey drift is a sandy to clayey calcareous drift. North of the central part of Lake Winnipeg and on the plains to the west, these drifts have yet to be distinguished and the former extent of Patrician ice is unknown. Farther south, some suggestion of persistent difference in heavy minerals is found in the work of Wallace and McCartney (1928), Kruger

(1937), Ehrlich and Rice (1955), and Dreimanis (1956). Amphiboles (hornblende) generally dominate in the grey drift whereas pyroxenes (augite) tend to dominate in the red. Of the iron oxides, haematite and limonite are more abundant in grey drift, and magnetite and ilmenite in red. Apatite and epidote are more common in grey drift and garnets and titanite more abundant in the grey. Chlorite, monazite and kyanite occur in grey drift but are not reported from the red.

Chemical analyses of the drifts by Ehrlich and Rice (1955), Grout and Soper (1914) and Ries and Keele (1912) are summarized:

|  | Percentage in grey drift | Percentage in red drift |
|---|---|---|
| Silica | 45–60 | 58–75 |
| Alumina | 6–10 | 11–15 |
| Iron oxide | 1–5 | 4–9 |
| Magnesia | 3–17 | 1.5–4 |
| Lime | 9–28 | 1.0–4 |
| Soda | 0.7–1.5 | 1.0–2.2 |
| Potash | 0.7–2.0 | 1.4–2.7 |
| Ignition loss | 2–15 | 2–6 |

The ignition loss represents combined water and carbon dioxide.

The red drift has a composition comparable with fresh rocks ranging from granite to diabase (Table III). The content of alumina, iron oxide, magnesia and potash is much lower than that of a deeply weathered rock, and the amounts of lime and soda much higher. The grey drift is a mixture of granitic rocks, carbonates and shale and its chemical composition superficially resembles that of some basic volcanic rocks.

*End moraines*

End moraines previously reported in the literature (references by Tyrrell, Hurst, Satterly, and Moorehouse) were traced by airphoto interpretation (Figure 3) by Elson (1957a, b) and Wilson *et al.* (1958), and by this method and field studies by Zoltai (1960) who also described the soils in them.

The area occupied by the end moraines is relatively small. They are described summarily as follows; numbers correspond to map designations on Figure 3:

1. Vermillion moraine (Upham, 1893; Leverett, 1932, p. 56): red drift, the west part extending underneath or incorporating grey drift, may correlate with the Darlingford-Edinburg-Holt system (*see* 10).

2. Eagle-Findlayson moraine (Moorehouse, 1940; Satterly, 1943; Antevs, 1951; Elson, 1957a; and Zoltai, 1960): stratified drift with a thin cover of till, locally till overlies lake deposits, top is flat, evidence of wave action on the sides.

3. Hartman moraine (Satterly, 1943, 1960): lithology similar to Eagle-Findlayson moraine, locally hummocky rather than flat-topped, may correlate with Dog Lake moraine farther east (Elson, 1957a), forms the northern limit of red lacustrine clay( Zoltai, 1960).

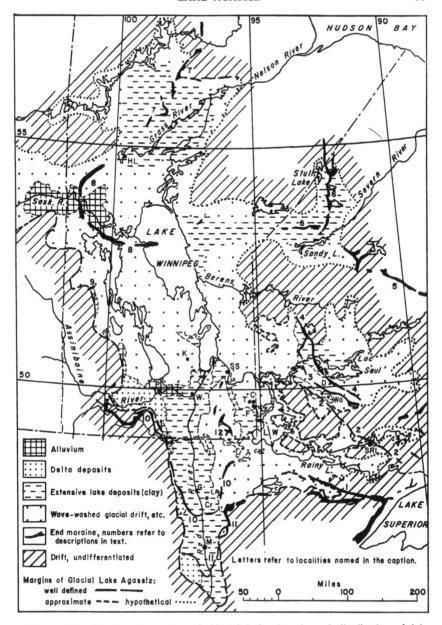

FIGURE 3. Map showing extent of Glacial Lake Agassiz and distribution of lake deposits and end moraines. Letters beside large dots indicate localities mentioned in the text as follows: C: Caddy Lake (Cross Lake); Cr: Crookston; D: Dryden; F: Falcon; G: Grand Forks; HL: Herb Lake; K: Komarno; LW: Lake of the Woods; M. Moorhead (Fargo); P: Portage la Prairie; SRL: Steep Rock Lake; SS: Seven Sisters; T: Telford; Wa: Wabigoon; W: Winnipeg.

4. Trout-Basket moraine (Hurst, 1932; Zoltai, 1960): lithology similar to Eagle-Findlayson, locally till overlies varved clay; may correlate with Kaiashk moraine farther east (Elson, 1957a).

5. Nipkip-Miminiska moraine (Camsell, 1904; Tyrrell, 1913 [Agutua moraine]; Satterly, 1490; Wilson *et al.*, 1958): lithology unknown, may contain some Palaeozoic carbonate rocks.

6. Sachigo moraine (Satterly, 1937; Wilson *et al.*, 1958): Precambrian lithology, southern end has wave-cut terraces, may be partly interlobate; a branch extends west north of Sandy Lake.

7. Burntwood-Etawney moraine system, mentioned vaguely in early reports but traced by Wilson *et al.* (1958): lithology unknown; may be partly interlobate. Separates west striations on east from southwest striations on west which seem to represent an earlier ice movement.

8. The Pas moraine (Dowling, 1901; Wilson *et al.*, 1958): calcareous till, flanked by beaches; till overlies lake deposits in the vicinity (Tyrrell, 1892).

9. Cowan moraine (Johnston, 1921): calcareous till, a short section of end moraine that marks the northern end of several of the higher Lake Agassiz strandlines (Elson, 1957b).

10. Darlingford-Edinburg-Holt moraine system (Leverett, 1932; Elson, 1958; Lemke and Colton, 1958): calcareous sandy till, may correlate with the Vermillion moraine (*see* 1).

11. Wahepton-Erskine moraine (Leverett, 1932): grey till, southernmost moraine in Lake Agassiz.

12. Whitemouth moraine (Johnston, 1921): calcareous till with beach deposits on surface, correlation unknown. Locally till overlies lake deposits (Elson, 1957a).

## Minor moraines

Washboard moraines are parallel ridges from 20 to 50 feet high, 100 to 200 feet wide, spaced about 500 feet apart that lie transverse to the direction of ice flow and form at or near ice margins standing in water (Mawdsley, 1936). Areas of these mapped on air photographs appear on the Glacial Map of Canada (Wilson *et al.*, 1958) in the eastern part of the Lake Agassiz basin. These have not been described from field examination, but such features generally consist of coarse gravelly till.

## Ice-contact stratified drift

A number of eskers appear within Lake Agassiz in western Ontario and northern Manitoba on the Glacial Map of Canada (Wilson *et al.*, 1958), but they are more abundant outside the area of the lake. Kame deposits occur at several places, notably northeast of Winnipeg (Birds Hill) and near the end moraines in western Ontario. These deposits are generally sand to pebble gravel with variable degrees of sorting.

### BASAL DEPOSITS IN LAKE AGASSIZ

In several localities in the Red River valley, thin sand and gravel bodies are reported between till and the overlying lake clay. These beds, known

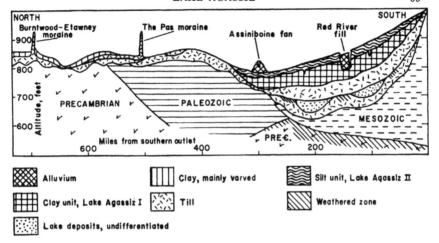

FIGURE 4.    Generalized schematic vertical cross section showing geology of the Lake Agassiz basin in a north-south belt west of Lake Winnipeg and along the axis of the Red River valley. The varved clay between the Burntwood-Etawney moraine and The Pas moraine may have been overridden by a glacial readvance.

only from borings, commonly yield artesian water and must have considerable lateral extent, probably in an east-west direction. They may have originated either as subaerial outwash or alluvial deposits, or as a result of currents and sorting in lake water 300 to 500 feet deep (Dennis, Akin, and Warts, 1949). No completely satisfactory mechanism for producing well sorted sediments at the base of a glacier standing in deep water is known to the writer.

Johnston (1915, p. 46) describes the gradual transition upward from till to laminated clays that takes place within eight or ten feet in the Rainy River district, and infers that the water was deep enough there to prevent disturbance of the bottom by wave action; stones dropped from floating ice. The resulting sediment has a grain-size distribution indistinguishable from till (compare curves L3 and L4 Figure 5a, and Figure 2).

## LAKE DEPOSITS

The "deep water" off-shore deposits of Lake Agassiz (Table I, Figures 3 and 4) are extensive but do not cover the floor of the whole basin. A large region in the latitude of Lake Winnipeg (Figure 3) has either no evidence of submergence or else patches of lag concentrate of boulders, sand, and gravel resting on till, or small, widely separated pockets of clay in protected basins. Part of the region may have been deglaciated under subaerial conditions, and part of it was remote from sources of sediment during submergence, so that the deep water deposits formed only a thin veneer that was later eroded.

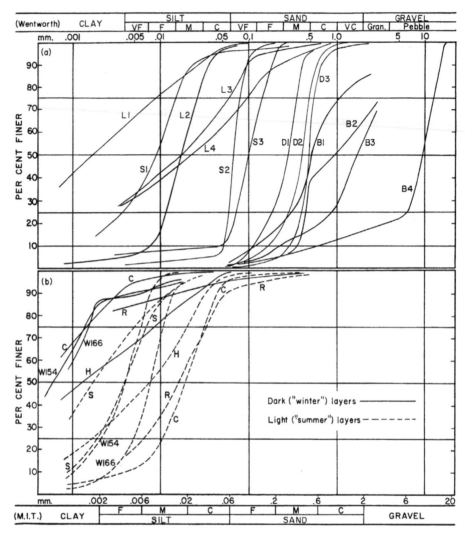

FIGURE 5a & b. Cumulative grain-size distribution curves of Lake Agassiz sediments. (a) Representative soils from Glacial Lake Agassiz: L1, L2, clay unit of Lake Agassiz I and basal part of silt unit of Lake Agassiz II at Winnipeg, after Wallace (1927); L3, L4, glacio-lacustrine clay, Rainy River area, after Johnston (1913); S1, silt from a pre-Agassiz glacial lake in the Assiniboine delta area (Elson, Geol. Surv. Canada, unpublished); S2, S3, littoral sand, Whitemouth and Rainy River areas after Johnston (1921, 1915); D1, D2, D3, deltaic sand from Assiniboine delta (Elson, Geol. Surv. Canada, unpublished); B1, B2, B3, beach sand and gravel from Hillsboro, Campbell, and McCauleyville beaches, after Laird (1944); B4, beach gravel from Roseisle, Man. (Elson, Geol. Surv. Canada, unpublished). (b) Varved clays: H. Herb Lake; C: Cross Lake (now Caddy Lake), after Wallace (1927); W154 and W166: Wabigoon, after Rittenhouse (1934); R: Rainy Lake, after Johnston (1915); S: Steep Rock Lake, after Legget and Bartley (1953).

In the Red River basin, the deep water sediments comprise a lower clay unit, deposited in the deep water of Lake Agassiz I, and an upper silt unit, deposited in the shallower water of Lake Agassiz II. The varved clays of western Ontario and northern Manitoba are presumed to have been formed in Lake Agassiz II.

### The clay unit of Lake Agassiz I

Dawson (1875, p. 248–50) recognized that an upper "fine yellowish marly and arenaceous clay" overlay a lower unit of laminated finer clay. This uniform lower clay extends from the latitude of Winnipeg, where it is twenty to forty feet thick, nearly to the south end of the basin. It is fifty to eighty-five feet thick in the central part of the Red River valley and lenses out laterally.

Baracos and Bozozuk (1958) divided the clay unit into an upper brown varved clay and a lower softer and siltier grey clay. The upper clay is thinly laminated. Wallace and Maynard (1924) found forty sedimentation planes in a thickness of one inch of clay at Winnipeg; but the laminations are not so thin everywhere. A grain-size analysis of the clay unit at Winnipeg (Wallace, 1927, Figure 1; this paper Figure 5a) showed a median grain size of about .0018 mm and about 38 per cent of the clay within the silt size range (Wentworth scale). Chemical analyses by Wallace and Maynard (1924) and Grout (1919) appear in Table V (Analyses 1 and 2). The high alumina and low potash content at Winnipeg suggests a mixture of granitic rocks with altered intermediate to basic lavas (Table III). According to Baracos and Bozozuk (1958), "the brown clay is approximately 30 per cent montmorillonite and the remainder practically all illite." This obviously refers to some fine fraction of the clay.

Fosness (1926) stated that the safe bearing capacity of this clay at Winnipeg is 1.5 to 2 tons per square foot. Baracos and Bozozuk (1958) found that swelling pressures of 1 ton per square foot are common and pressures up to 20 tons per square foot can be generated by air-dried samples.

Rominger and Rutledge (1952) presented data on soil mechanics for the Grand Forks-Crookston-Fargo region. The upper part of the clay unit there generally has liquid limits of 100 to 120 per cent and plasticity indices of 60 to 80 per cent. The lower part has liquid limits of 75 to 85 per cent and plasticity indices of 40 to 60 per cent. The surface of the upper part of the clay unit has been preconsolidated by desiccation. They do not describe the lithology in terms similar to those of Baracos and Bozozuk, but the data on soil mechanics are compatible with the lithology at Winnipeg.

Both parts of the clay unit were formed in Lake Agassiz I. The lower silty massive clay may have been deposited either near the ice margin retreating in deep water or else in water shallower than the wave base some distance from the glacier. The upper laminated clay was laid down in deep water when the ice margin was remote.

## TABLE V
### CHEMICAL ANALYSES OF CLAYS

| | Lake Agassiz I | | | | | Lake Agassiz II | | | | | Average | Alluvium | Early lake |
|---|---|---|---|---|---|---|---|---|---|---|---|---|---|
| | 1 | 2 | 3 | 4 | 5 | 6 | 7 | 8 | 9 | 10 | 11 | 12 | 13 |
| Silica | 55.48 | 47.70 | 56.02 | 58.63 | 53.32 | 69.38 | 54.65 | 49.86 | 50.33 | 54.48 | 55.83 | 54.00 | 45.15 |
| Alumina | 20.37 | 13.58 | 20.53 | 9.12 | 8.87 | 15.76 | 20.94 | 16.97 | 17.86 | 16.09 | 15.77 | 9.25 | 9.05 |
| Iron oxides | 5.54 | 6.51 | 5.35 | 8.42 | 4.71 | .90 | 6.94 | 7.41 | 8.17 | 5.99 | 6.36 | 2.77 | 3.75 |
| Magnesia | 3.29 | 3.13 | 3.81 | 3.62 | 6.62 | .85 | 2.69 | 4.64 | 4.66 | 3.86 | 3.84 | 3.51 | 7.11 |
| Lime | 4.95 | 11.70 | 2.75 | 6.41 | 9.21 | 4.38 | 2.84 | 8.41 | 6.37 | 5.50 | 5.73 | 9.77 | 14.00 |
| Soda | .82 | 1.05 | 2.80 | 1.60 | 6.60 | 2.10 | 1.20 | 1.01 | 1.12 | 2.59 | 2.13 | | |
| Potash | 1.86 | 2.34 | 2.22 | 2.85 | 2.26 | 2.11 | 2.29 | 2.66 | 2.87 | 2.85 | 2.51 | 2.34 | 2.52 |
| Moisture | | 1.79 | | | 1.94 | | | | | | | 8.66 | 1.50 |
| Ignition loss | 8.77 | 13.23 | 9.02 | 9.00 | 6.07 | 2.00 | 8.16 | 9.97 | 8.63 | 9.34 | 8.02 | 9.95 | 16.82 |
| Other | | .31 | | | .37 | | | | | | | .05 | .10 |
| | 100.08 | 101.34 | 100.50 | 99.53 | 99.97 | 100.48 | 99.71 | 100.93 | 100.11 | 100.43 | 100.19 | — | — |

EXPLANATION:
1. Winnipeg, clay unit of Lake Agassiz I, analysed by Wallace and Maynard (1924).
2. Net Lake Rapids, Minn., grey clay, Grout (1919).
3 and 4. Winnipeg, silt unit, upper grey clay and lower sandy clay, Wallace and Maynard (1924).
5. East Grand Forks, Minn., silt unit of Lake Agassiz II, Grout (1919).
6 and 7. Cross Lake (Caddy Lake), Man., varved clay, summer and winter layers, Wallace (1927).
8 and 9. Herb Lake, Man., varved clay, summer and winter layers, average of three each, after Wallace (1927).
10. Dryden, Ont., varved clay, red, bulk analysis, Keele (1924).
11. Lake Agassiz II: average of analyses 3–10.
12. Portage la Prairie, Man., alluvium, Ries and Keele (1912).
13. Gilbert Plains, Man., clay of a small lake ponded against the Manitoba Escarpment just prior to the formation of Lake Agassiz I, Ries and Keele (1912).

*The silt unit of Lake Agassiz II*

Silty clay now forming much of the floor of the Lake Agassiz basin was deposited in Lake Agassiz II. In western Ontario and northern Manitoba within roughly forty to eighty miles of the ice margin varved clays (laminated silts and clays) formed. In the Red River valley, more than 200 miles from the ice margin the sediments were derived partly from rivers tributary to the lake. Here laminations are poorly developed or absent.

The silt unit rests unconformably on the undulating drying surface of the clay unit. Locally sand and gravel form the base of the silt unit, and in places the sand contains mollusc shells, peat, and wood. Wood from Moorhead has been dated by radiocarbon methods at the United States Geological Survey (Rubin and Alexander, 1958, W388) as about 9,900 years old. The silt unit is much less uniform than the underlying clay unit. Its composition ranges from silt to clay and inclusions of sand, silt or clay are common. In places it is sufficiently permeable to function as an aquifer. The colour is generally yellow to grey. At Winnipeg, the silt unit is 8 to 15 feet thick; in the central part of the Red River basin it is 15 to 35 feet thick and the upper beds of the unit are laminated. No grain-size or mineralogical analyses of the silt unit as such are known to the writer. Wallace and Maynard (1924) gave two chemical analyses of the silt unit at Winnipeg, and Grout (1919) one at East Grand Forks, Minnesota (Table V).

In Winnipeg, the mechanical properties of the silt unit create serious foundation problems. At its base, generally 8 to 13 feet below the surface, is a sandy-silty yellow clay bed 0.5 to 2 feet thick with a safe bearing capacity of only 0.5 ton per square foot or less. The overlying silt is about twice as strong, but most buildings of even moderate size must be founded on the brown clay of Lake Agassiz I. In the southern part of the basin, at Grand Forks and Fargo, Rominger and Rutledge (1952) found that the liquid limit of the silt unit is generally between 60 and 70 per cent and the plasticity index about 40 per cent. The natural water content, 30 to 50 per cent, is greatest near the base.

*Varved clays*

The lake sediments in the east and north parts of the basin were formed in Lake Agassiz II. Their well-developed lamination has been studied by Wallace (1927), Antevs (1931, 1951), Rittenhouse (1934), Eden (1955) and others. Most exposures are less than a dozen feet thick, but locally the varved clays are fifty feet or more thick. They rest on bedrock, till or thin stratified drift deposits overlying till. In northern and southeastern Manitoba, extensive clay deposits are covered with peat and form broad, marshy plains.

The number of varves in any locality varies with the history of that part of the lake. The results of several studies of varves follow:

Antevs (1931), Grass River region, Manitoba, upper sequence of about 375 varves was correlated, lower thinly laminated sequence estimated at more than 700 varves.

Antevs (1951), Steep Rock Lake, Ontario, eight distinct series totalling about
    825 varves.
Satterly (1937, 1938a), Stull Lake, Sachigo Lake, Ontario, about 280 varves.
Satterly (1938b), Sandy Lake, Ontario, about 240 varves.

As may be expected, the varve sequences are thinner and contain fewer
couplets in the north than they do in the south.

The grain size of the light and dark layers (Figure 5b) was studied by
Johnston (1915, p. 45), Wallace (1927) and Rittenhouse (1934). The
median grain size of the dark ("winter") layer is roughly .0005 to .0008 mm
and may be as large as .0013 mm; from 8 to 16 per cent of the dark layer
is silt. The dark layers at Herb Lake in northern Manitoba are coarser than
others studied, comprising about 36 per cent silt. The light ("summer")
layers are mainly silt and show more variation with median grain sizes
ranging from .0016 to .02 mm. They contain from 10 to 60 per cent clay,
but generally less than 40 per cent, and as much as 10 per cent sand. Suther-
land (Legget and Bartley, 1953) found that the dark laminae at Steep
Rock Lake contained 50 per cent by weight of grains smaller than .002 mm
whereas the light laminae contained only 10 per cent. Eden (1955)
studied the grain size within a dark layer about three inches thick and
found that it was essentially homogenous (symminct). By comparison,
varves from the Don Valley Interglacial at Toronto showed a continuous
decrease in grain size from the bottom of the light layer to the top of the
dark layer (diatactic structure).

Eden (1955) found that the light layers at Steep Rock Lake have
liquid limits of from 23 to 27 per cent, plasticity indices of 3 to 5.5 per cent
and natural water contents a per cent or two higher than the liquid limit.
Their unstable state is a major factor contributing to the mass movement of
clays in the now drained basin of Steep Rock Lake. The dark layers have
liquid limits generally ranging from 75 to 95 per cent, plastic indices of 40 to
60 per cent, and natural water contents a per cent or two below the liquid
limit. Both layers are extrasensitive soils. At the Seven Sisters power site,
about 50 miles east-northeast of Winnipeg, Peterson et al. (1958) found that
the upper 10 to 20 feet of highly plastic clays have a liquid limit of about
85 per cent, plastic limit 26 per cent, and natural water content 45 per cent;
an underlying thinner clay of lower plasticity, has a liquid limit of 31 per
cent, plastic limit of 13 per cent, and water content of 21 per cent.

Clay from the Seven Sisters project was analysed by Lambe and Martin
(1956); which of the clays is not specified, but the Atterberg limits suggest
the upper highly plastic clay. The composition by weight of the portion
finer than .074 mm is listed as:

|                  |              |
|------------------|--------------|
| Illite           | 45 per cent  |
| Montmorillonoid  | 40           |
| Calcite          | 3            |
| Organic matter   | 2            |

Wallace (1927) made chemical analyses of individual varves and a bulk analysis of clay at Dryden is reported by Keele (1924) (Table V). Ritten-house (1934) analysed Wabigoon clays for iron and carbonate and found the dark layers contained twice as much iron as the light layers; however, his "abnormal" dark layers (red) had about the same iron content as "normal" dark layers (grey) and he concluded that red represented the ferric state and grey the ferrous state. The thick red layers ("abnormal dark") contained about 60 per cent more carbonate than the thinner grey ("normal dark") layers, but Rittenhouse found no significant difference in the carbonate in light and dark layers of ordinary varves. In all of the analyses presented by Wallace, the carbonate content was appreciably higher in light ("summer") layers and iron and alumina were consistently higher in the dark ("winter") layers (Table V).

Wallace's Cross Lake (now Caddy Lake, on the Canadian Pacific Railway three miles west of the Manitoba-Ontario boundary) dark ("winter") layers have a chemical composition similar to that of the clays of roughly the same age at Winnipeg about 80 miles west, which suggests that the fine fraction of the clays was dispersed throughout Lake Agassiz whereas the coarser fraction remained close to its source. This also indicates that the material that formed the clays came mainly from the ice margin rather than the rivers entering the opposite side of the lake. (Compare with Portage la Prairie clay (Table V), alluvium derived from the west side of the Lake Agassiz basin.)

Theories of varve deposition are summarized by Eden (1955) and Legget and Bartley (1953). Doubt has been expressed by the latter authors and Deane (1950) as to the annual periodicity of varves, without necessarily challenging the validity of the laborious and meticulous work of Antevs and others in correlating individual varve sequences.

The sediment that forms varves probably spreads out in the bottom tens of feet of large glacial lakes as density currents. Silt forming light layers settles out rapidly but the clayey material of the dark layers is precipitated by geochemical changes in the lake that cause flocculation. Thus, the fine sediment may be widely dispersed and the thickness of the dark layers may represent the amount of glacial melting, the distance from a source, or the trapping of sediment in a local basin. The light layer (silt) may be subject to similar controls, but because its particles are less subject to flocculation the grain size is more directly related to the distance from the source. The grain size of the light layers should be a useful parameter in varve correlation.

### Red clay

Red clay layers, studied first by Rittenhouse (1934), at Wabigoon, were reported by Antevs (1951, p. 1238) at Steep Rock Lake. At Wabigoon the red clays are abnormally thick and are overlain by about 360 varves; at Steep Rock Lake they are overlain by 176 varves. Other occurrences of laminated red clay in similar positions are reported by Keele (1924, p. 153) near Shoal Lake, and Zoltai (1960) in the Turtle River basin, and generally southwest of the Hartman moraine. Zoltai reports that near Rainy Lake and Lake of the Woods the red clay is massive rather than varved. If these are all the same bed, the laminations evidently become imperceptible in a distance of about 40 miles.

Gleeson (1960) analysed a red clay bed from 0.5 to 2 feet thick underlying 5 to 10 feet of grey clay, and overlying about 30 feet of grey clay in Blacky Bay of Kakagi Lake just east of Lake of the Woods. He found that the clays are a "rock flour" composed of quartz, feldspar, mica, chlorite, amphibole, some "clay" chlorite, mixed layers, mica minerals, and hydrobiotite. Chemical analysis showed that the red clay is relatively rich in iron, lime and magnesia, whereas the grey clay has more silica and soda. The red clay has a "gabbroic" composition, not unlike the grey drift, and the grey clay a "dioritic" composition, which suggests some difference in provenance. The red colour seems to be due to the ferric state of the iron rather than excessive amounts of iron. If the environment that produced it was widespread and of short duration the red clay may be a useful marker horizon.

*Intersecting minor ridges*

Horberg (1951) described low, intersecting ridges 3 to 10 feet high and 75 to 500 feet wide which form a "fracture" pattern on lake sediments in the Red River valley. This pattern is not discernible on the ground but is conspicuous in air photographs. Similar patterns occur on the floors of other glacial lakes. The ridges are mainly in the central part of the basin and cross several of the youngest lake Agassiz beaches. Horberg recognized six types of patterns including polygonal ridges, cellular networks, cuneiform ridges, furrow polygons, furrow networks and cuneiform furrows. He compared these with similar patterns in areas of permanently frozen ground and, influenced by wedge structures and involutions visible in shallow road cuts, concluded that the ridges represent frozen-ground structures formed during the retreat of the late Wisconsin ice.

Nikiforoff (1952) thought the involutions and wedge structures were formed by other than permanently frozen ground processes and that the ridges were due to wave action in the final, shallow phase of the lake. Colton (1958), after studying similar features in Montana, noticed crosscutting among some Lake Agassiz ridges, and that some ridges were superimposed on others. He postulated that the pattern formed when fluctuations of lake-level lowered several feet of lake ice, through which long narrow open-water leads extended, onto the lake bottom, and the weight of the ice forced the soft sediments up into the openings through the ice.

Patterns of this sort are being recognized in many types of surficial deposits, and can be related to fracture patterns in the underlying bedrock. Minute crustal displacements are manifest through considerable thicknesses of soil as loci of streams, gullies, lake margins, ridges and depressions (Russell, 1958). The relationship is not yet understood.

## Near-shore deposits

In the south and west parts of Lake Agassiz deposits of very fine sand or silt (Figure 5a) with varying amounts of sand and clay, mostly less than twenty feet thick, occur in a belt six to ten miles wide which extends down from about the 850 foot contour. The silt lenses out on the "shore" side and grades into clay on the "lake" side. It represents fine material winnowed by wave action from the till forming the shores.

## Shore deposits

Shores of Lake Agassiz are marked by wave cut scarps, terraces, and beach ridges. Several of the beach ridges can be traced for hundreds of miles and form the basis for studies of post-glacial crustal warping (Johnston, 1946). Beach ridges are sand and gravel deposits from 2 to 20 feet high and 100 to 300 feet wide. Commonly the sand and gravel is five to eight feet thick in the centre, and forms a body with a lenticular cross section. The grain size of the sand and gravel depends on the material attacked by the waves. Laird (1944, p. 8) noted that certain beaches appear to have characteristic grain-size distributions. Some beach deposits show remarkable sorting (Figure 5a) and bimodal grain-size distributions are common. Where thicker than ten feet the beach ridges may serve as aquifers for domestic wells.

Development of beach ridges depends on the slope of the shore as well as the type of material available. In southwestern Manitoba, beach ridges are best developed on slopes of fifty to sixty feet per mile (about 1 in 100). On slopes steeper than eighty feet per mile (about 1.5 in 100) the sand and gravel are moved to deep water and shore scarps form; on slopes of less than thirty feet per mile (about 0.6 in 100) waves are impeded by bottom drag and only weak features result.

Between many beaches, and at the foot of scarps, are wave-cut terraces covered by lag concentrates comprising boulders up to several feet in diameter and coarse gravel. In places the boulders are so numerous that they completely cover the underlying till. A rough estimate of the quantity of boulders in the till and the residual sand and gravel now in the form of beach ridges indicates that twenty to forty feet of drift must have been eroded by the waves to produce many of these features.

## Deltas

Four large deltas were built into Lake Agassiz I by tributary streams from the west; normal run-off may have been augmented by glacial meltwater during their deposition. From south to north they are the Sheyenne, Elk Valley, Pembina, and Assiniboine deltas. The last is the largest. The upper surfaces of these deltas correlate with the highest strandlines. The northeast sides of the deltas may have been partly controlled by the glacier margin (Figure 3). Except for the Elk Valley delta which received direct contributions from the glacier, the delta sediments can be accounted for by erosion of the valleys of the streams concerned. The deltas are gravel and coarse sand at their apexes, and medium to coarse sand (Figure 5a) over much of their surfaces (topset beds). The distal portions are silt and very fine sand. Medium to coarse sand is susceptible to movement by the wind and forms extensive areas of sand dunes. The delta sediments range from a few feet to more than 250 feet in thickness, and overlap deep-water clay and silt.

Secondary delta deposits form paired terraces in part of the Assiniboine valley which became an estuary of Lake Agassiz II. From twenty to fifty feet of medium to coarse sand are overlain by ten to twenty feet of clay and silt. Unlike the deposits of the first lake these beds contain fossils (molluscs). Several small valleys, eroded in the older Assiniboine delta, were partly filled with lake sediments and alluvium at this time.

## POST-GLACIAL AND RECENT DEPOSITS

### Aeolian sand

Windblown sands have already been mentioned in connection with delta deposits. Most are on the west side of Lake Agassiz. In a few places dunes are localized along former strandlines where wave action has concentrated sand of medium to coarse grain. Capillary action in fine sands tends to keep them moist so that they are well anchored by surface tension and by vegetation and do not blow unless disturbed by cultivation. Most dune areas are now stable. The morphology of the dunes indicates that they were formed by a wind pattern essentially the same as the present one.

### Alluvium

Alluvium is accumulating on flood plains of many rivers and on alluvial fans at the foot of the Manitoba Escarpment. Flood plain alluvium is generally silt and sand with variable amounts of clay. Alluvium in the fans ranges from clay to pebble gravel.

The alluvium of the Red River resembles some deposits of Lake Agassiz. It forms terraces about ten feet below the lake plain and as much as twenty-five feet above river level (Horberg, 1951). The Red River eroded a valley somewhat deeper than the present one after the draining of Lake Agassiz, roughly 7,000 years ago. Subsequently (Rubin and Alexander, 1960, p. 175; W860 and W862, 6200 and 6750 years respectively) a drier climate reduced the discharge of the river and, perhaps in conjunction with a reduction of gradient due to crustal warping, caused the river to adjust to a smaller channel and partly fill the larger valley. Alluvium, mainly silty clay, is accumulating on relatively narrow terraces within the original valley (Figure 4). It contains fossils and archaeological material dating almost to historical times north of Winnipeg (MacNeish, 1958).

The type of sediment in the alluvial fans depends on the nature of the rock forming the Manitoba Escarpment. South of the Assiniboine River many of the east-flowing streams have cut through the siliceous, pebble-forming Odanah (Riding Mountain) shale and are eroding the underlying clay shales (Millwood shale, Vermillion River formation) which weather to massive clay. In this region sand and shale gravel in alluvial fans is irregularly interbedded with clay. In places fans cover beach gravels and wave-cut terraces with a thin veneer of clay. North of the Assiniboine River the stream channels are steeper and are eroding Odanah shale. Here alluvial fans composed of pebble gravel and sand cover old beach ridges and wave-washed drift along the base of Riding, Duck and undoubtedly Porcupine mountains and the Pasquia Hills. They occur in a belt about four to ten miles wide (Johnston, 1934) and some are as thick as forty feet.

Extensive alluvial fans that extend into lakes to form deltas are being built by Saskatchewan and Assiniboine rivers. The alluvium of the Saskatchewan River (Kuiper, 1954) has not been well described but is presumably similar to that of the Assiniboine River but sandier. The Assiniboine alluvial fan comprises sandy channel deposits and clay fillings of abandoned channels flanked by the silty material of natural levees and clayey backswamp deposits. These typical overbank flood plain sediments are superimposed to form a considerable thickness of alluvium. The Assiniboine River has filled what was originally the south end of Lake Manitoba. At Portage la Prairie clayey alluvium is used for brick making (Table V).

GROUNDWATER

Groundwater acquires from the soil characteristics that govern its effects on pipelines and on structural materials such as concrete and steel, as well as its usefulness as a water supply. Table VI, showing the chemical nature of groundwater in the region near the International Boundary, is based on

TABLE VI

CHEMICAL CONTENT OF WATERS FROM SURFACE DEPOSITS OF THE LAKE AGASSIZ BASIN
After Allison, Johnston, and Paulson
Range of values in parts per million

|  | Red drift | Grey drift | Lake Agassiz silt unit | Salt Spring (Wallace) |
|---|---|---|---|---|
| Ca | 55–60 | 50–430 | 20–448 | 2,000 |
| Mg | 15–25 | 20–140 | 120–440 | 550 |
| Na | 15–60 | 60–220 | 45–160 | 35,000 |
| K | 5–15 | 5–35 | 30 | 1,400 |
| $HCO_3$ | 150–320 | 210–470 | 440–740 | 200 |
| $SO_4$ | 10–65 | 5–670 | 250–1,850 | 4,900 |
| Cl | 5–25 | 3–1,350 | 5–250 | 55,950 |
| Fe | .5–4 | 0–42 | 0–2 |  |
| Total dissolved solids | 250–450 | 30–2,000 | 1,100–3,500 |  |
| Total hardness | 220–360 | 200–1,650 | 700–2,300 |  |

water supply papers by Allison (1932), Thiel (1947), Paulson (1951), and Johnston (1934). Lower precipitation, with reduced leaching, as well as a more soluble drift lithology including gypsum, make the groundwater on the west side of the basin much richer in dissolved salts; the high values in Table VI generally pertain to the western area.

Water in red drift is neutral or slightly acid; it is soft compared with waters of the grey drift, though usually classed as a hard water. Waters from sand and gravel contained in grey drift are very hard and contain moderate to large amounts of the bicarbonate, sulphate, and chloride radicles. The clay unit of Lake Agassiz I is too impermeable to create groundwater problems. Water in the more permeable silt unit of Lake Agassiz II is a serious threat to buried pipes, etc. (Sill, 1953). It is extremely hard and has moderate to very high amounts of bicarbonate and sulphate radicles, and moderate amounts of chlorine.

A number of springs emerge through the surface deposits along the foot of the Manitoba Escarpment, their source apparently being the lower Cretaceous sandstone. Some of their water passes through Palaeozoic rocks before reaching the surface as in the area west of Lake Winnipegosis. Wallace (1917) described the highly corrosive effect of these salt springs on boulders in the drift. Presumably similar water permeates the drift elsewhere without reaching the surface as springs.

## Conclusion

The foregoing discussion is a synthesis of available information, unfortunately only skeletal, from widely scattered studies with a variety of purposes, some practical, some academic, and some incidental to other objectives. The viewpoints of geology, engineering, agronomy and industry have been focused on a mutual problem. It is hoped that this paper will suggest new approaches to students of glacial lakes in general, and of Glacial Lake Agassiz in particular.

## ACKNOWLEDGEMENTS

The author is grateful to Stephen C. Zoltai for permission to refer to his unpublished manuscript on western Ontario. Several colleagues have suggested much appreciated sources of information. Jeanne B. Elson typed and edited the manuscript; without her generous assistance its inclusion in this volume would not have been possible.

## REFERENCES

Lake deposits in Manitoba were recorded by Dawson (1875), Dowling (1896, 1901), Upham (1890), Tyrrell (1890), Tyrrell and Dowling (1900), McInnes (1913), Wright (1925, 1927), and Johnston (1916, 1917, 1918, 1921, 1934, 1946), Quinn (1954), and Mulligan (1955). Lake deposits, here presumed to be those of Lake Agassiz, were reported in Ontario by Dowling (1896), Wilson (1909), Bell (1891), and Camsell (1904), whose reports were republished by Miller (1912); and Johnston (1915, 1917, 1946), Keele (1924), Bruce and Hawley (1927), Greig (1927), Hurst (1930, 1933), Burwash (1933), Satterly (1937, 1938a, b, 1940, 1943), Moorehouse (1940), and Zoltai (1960). Not all these writers connected the deposits they observed with Lake Agassiz and the writer is responsible for some correlations. The United States portion of Lake Agassiz was mapped in whole or in part by Upham (1895), Hall (1905), Leverett (1932), Laird (1944), Dennis, with various others (1949, 2 papers; 1950, 2 papers), and Paulson (1951, 1953).

1. ALLAIRE, A. (1916). The failure and righting of a million-bushel grain elevator. Amer. Soc. Civil Engineers, Trans., *80*: 799–843.

2. ALLISON, I. S. (1932). The geology and water resources of northwestern Minnesota. Minn. Geol. Surv., Bull. 22.

3. ANTEVS, ERNST (1931). Late-glacial correlations and ice recession in Manitoba. Geol. Surv., Can., Mem. 168.

4. ———— (1951). Glacial clays in Steep Rock Lake, Ontario, Canada. Geol. Soc. Amer. Bull., *62*: 1223–62.

5. BARACOS, A. (1952). Foundation investigation in Winnipeg following the 1950 Red River flood. Nat. Res. Council of Canada, Assoc. Comm. on Soil and Snow Mechanics, Proceed. Fifth Can. Soil Mechanics Conference, January 1952, Tech. Mem. 23: 56–8.

6. ———— and BOZOZUK, M. (1958). Seasonal movements in some Canadian clays. Proc. Fourth Internat. Conf. Soil Mechanics and Foundation Engineering, London, August 1957, 264–8.

7. ———— and MARANTZ, O. (1953). Vertical ground movements. Nat. Res. Council of Canada, Assoc. Comm. on Soil and Snow Mechanics, Proceed. Sixth Can. Soil Mechanics Conference, Winnipeg, Dec. 1952, Tech. Mem. 27: 29–36.

8. BATEMAN, J. D. (1940). Geology and gold deposits of the Uchi–Slate Lakes area. Ont. Dept. Mines, *48*, pt. VIII, 1939: 1–43.

9. BELL, ROBERT (1891). Report on an exploration of portions of the Attawapiskat and Albany Rivers Lonely Lake to James Bay, 2, pt. G. Geol. Surv., Can.; also in Ontario Bureau of Mines, *21*, pt. 2: 59–83.

10. BRUCE, E. L., and HAWLEY, J. E. (1927). Geology of the basin of Red Lake, District of Kenora. Ont. Dept. Mines, *36*, pt. III: 48–9.

11. BURWASH, E. M. (1933). Geology of the Kakagi Lake area. Ont. Dept. Mines, *42*, pt. IV: 71–4.

12. CAMSELL, CHARLES (1904). Country around headwaters of Severn River. Geol. Surv., Can., *16*, pt. A: 143–50; also in Ont. Bureau Mines, *21*, pt. 2: 87–93.

13. COLTON, R. B. (1958). Note on the intersecting minor ridges in the Lake Agassiz basin, North Dakota. *In* Guidebook, Ninth Annual Field Conference, Mid-western Friends of the Pleistocene, May 17–18, 1958, North Dakota Geol. Surv., Misc. Ser. no. 10: 74–77.

14. DAWSON, G. M. (1875). Report on the geology and resources of the region in the vicinity of the Forty-ninth Parallel, from the Lake of the Woods to the Rocky Mountains, Montreal.

15. DEANE, R. E. (1950). Pleistocene geology of the Lake Simcoe district, Ontario. Geol. Surv., Can., Mem. 256.

16. DENNIS, P. E., and AKIN, P. D. (1950). Ground water in the Portland area, Traill County, North Dakota. North Dakota Geol. Surv., ground water studies, no. 15.

17. ————, ————, and JONES, S. L. (1949). Ground water in the Wyndmere area, Richland County, North Dakota. U.S. Geol. Surv. and North Dakota Geol. Surv., North Dakota ground water studies, no. 13.

18. ————, ————, and ———— (1950). Ground water in the Kindred area, Cass and Richland Counties, North Dakota. U.S. Geol. Surv. and North Dakota Geol. Surv., North Dakota ground water studies, no. 14.

19. ————, ————, and WORTS, G. F., Jr. (1949). Geology and ground-water resources of parts of Cass and Clay counties, North Dakota and Minnesota. U.S. Geol. Surv. and North Dakota Geol. Surv., North Dakota ground water studies, no. 10: 17–29.

20. Dowling, D. B. (1896). Red Lake and part of the basin of Berens River, Keewatin. Geol. Surv., Can., Ann. Rept., vol. VII, pt. F: 51–54; also in Ont. Bureau Mines, 21, pt. 2: 19–48.

21. —— (1901). Report on the geology of the west shore and islands of Lake Winnipeg. Geol. Surv., Can., Ann. Rept., vol. XI, pt. F.

22. Dreimanis, Aleksis (1956). Steep Rock iron ore boulder train. Geol. Assoc. Canada, Proceed., 8, pt. 1: 27–70.

23. Eden, W. J. (1955). A laboratory study of varved clay from Steep Rock Lake, Ontario. Amer. J. Sci., 253: 659–74.

24. Ehrlich, W. A., and Rice, H. M. (1955). Postglacial weathering of Mankato till in Manitoba. J. Geol., 63: 527–37.

25. Ellis, J. H., and Pratt, L. E. (1953). Source and nature of regolith in the various landscape areas of southern Manitoba. Nat. Res. Council of Canada, Assoc. Comm. on Soil and Snow Mechanics, Proceed. Sixth Can. Soil Mechanics Conference, Winnipeg, Dec. 1952, Tech. Mem. 27, 3–20.

26. ——, and Shafer, W. H. (1943). Reconnaissance soil survey of south-central Manitoba. Manitoba Dept. Agric., Soils Rept. no. 4.

27. Elson, J. A. (1957a). Lake Agassiz and the Mankato-Valders problem. Science, 126: 999–1002.

28. —— (1957b). History of glacial Lake Agassiz. Manuscript presented at the Fifth Congress of the International Association on Quaternary Research, Madrid, 1957; to be published in V Actas Inqua (processed).

29. —— (1958). Pleistocene history of southwestern Manitoba. In Guidebook, Ninth annual field conference, Mid-western Friends of the Pleistocene, May 17–18, 1958, North Dakota Geol. Surv., Misc. Ser. no. 10, 62–73.

30. Flint, R. F. (chairman) et al. (1959). Glacial Map of the United States east of the Rocky Mountains. Geol. Soc. Amer.

31. Fosness, A. W. (1926). Foundations in the Winnipeg area. Engineering Journal, vol. IX, no. 12, Dec. 1926, 495–503.

32. Gleeson, C. F. (1960). Studies on the distribution of metals in bogs and glaciolacustrine deposits. Unpublished Ph.D. thesis, McGill University.

33. Greig, J. W. (1927). Woman and Narrow Lakes area, District of Kenora (Patricia Portion). Ont. Dept. Mines, 36, pt. 3: 85–110.

34. Grout, F. G. (1919). Clays and shales of Minnesota. U.S. Geol. Surv., Bull. 678.

35. ——, and Soper, E. K. (1914). Preliminary report on the clays and shales of Minnesota. Minn. Geol. Surv., Bull. 11.

36. Hall, C. M. (1905). Casselton and Fargo quadrangles. U. S. Geol. Surv., Folio no. 117.

37. Hills, G. A., and Morwick, F. F. (1944). Reconnaissance soil survey of parts of northwestern Ontario. Ontario Soil Survey, Rept. 8.

38. Horberg, Leland (1951). Intersecting minor ridges and periglacial features in the Lake Agassiz basin, North Dakota. J. Geol., 59: 1–18.

39. Hurst, M. E. (1930). Geology of the area between Favourable Lake and Sandy Lake, District of Kenora (Patricia Portion). Ont. Dept. Mines, 38, pt. ii: 67–8.

40. —— (1933). Geology of the Sioux Lookout area. Ont. Dept. Mines, 41, pt. vi: 16–18.

41. Johnston, W. A. (1915). Rainy River district, Ontario, surficial geology and soils. Geol. Surv., Can., Mem. 82.

42. —— (1916). The genesis of Lake Agassiz; a confirmation. J. Geol., 24: 625–38.

43. —— (1917). Records of Lake Agassiz in southeastern Manitoba and adjacent parts of Ontario, Canada. Geol. Soc. Amer. Bull., 28: 145–8.

44. —— (1918). Reconnaissance soil survey of the area along the Hudson Bay Railway. Geol. Surv., Can., Summ. Rept. 1917, pt. D, 30–6.

45. —— (1921). Winnipegosis and Upper Whitemouth River areas, Manitoba, Pleistocene and Recent deposits. Geol. Surv., Can., Mem. 128.

46. ——— (1934). Surface deposits and ground-water supply of the Winnipeg map-area, Manitoba. Geol. Surv., Can., Mem. 174.

47. ——— (1935). Patrician center of glaciation. Pan-Am Geologist, 63, no. 1: 13–18.

48. ——— (1946). Glacial Lake Agassiz, with special reference to the mode of deformation of the beaches. Geol. Surv., Can., Bull. 7.

49. KEELE, JOSEPH (1924). Preliminary report on the clay and shale deposits of Ontario. Geol. Surv., Can., Mem. 142, 125–33.

50. KRUGER, F. C. (1937). A sedimentary and petrographic study of certain glacial drifts of Minnesota. Amer. J. Sci., 234: 345–63.

51. KUIPER, EDWARD (1954). Interim report no. 8, Saskatchewan River Reclamation Project, Hydrometric surveys 1954. Canada, Dept. Agric., Prairie Farm Rehabilitation Administration, Engineering Branch (processed).

52. LAIRD, W. M. (1944). The geology and ground water resources of the Emerado Quadrangle. North Dakota Geol. Surv., Bull. 17.

53. LAMBE, W. T., and MARTIN, R. T. (1956). Composition and engineering properties of soil (IV). Nat. Acad. Sci.-Nat. Res. Council, Highway Research Board, Proceed. Thirty-fifth Ann. Meeting, Wash., D.C., Jan. 17–20, 1956, 661–77 (p. 672).

54. LAWSON, A. C. (1885). Report on the Geology of Lake of the Woods Region. Geol. and Nat. Hist. Surv., Can., Ann. Rept. new ser. 1, pt. CC: 130cc–140cc.

55. LEGGET, R. F., and BARTLEY, M. W. (1953). An engineering study of glacial deposits at Steep Rock Lake, Ontario, Canada. Econ. Geol., 48: 513–40.

56. LEMKE, R. W., and COLTON, R. B. (1958). Summary of the Pleistocene geology of North Dakota. In Guidebook, ninth annual field conference, Mid-western Friends of the Pleistocene, May 17–18, 1958, North Dakota Geol. Surv., Misc. Ser. no. 10, 41–57.

57. LEVERETT, FRANK (1932). Quaternary geology of Minnesota and parts of adjacent states. U. S. Geol. Surv., Prof. Pap. 161.

58. MACDONALD, A. E. (1937). Report of the Winnipeg Branch of the Engineering Institute of Canada Committee on Foundations. Engineering Journal, XX, no. 11, Nov. 1937, 827–9.

59. MCINNES, WILLIAM (1913). The basins of Nelson and Churchill rivers. Geol. Surv., Can., Mem. 30, 118–27.

60. MACNEISH, R. S. (1958). An introduction to the archaeology of southeast Manitoba. Nat. Mus. Canada, Bull. 157.

61. MANZ, O. E. (1956). Investigation of Lake Agassiz clay deposits. North Dakota Geol. Surv., Rept. Inv. 27.

62. MAWDSLEY, J. B. (1936). The wash-board moraines of the Opawica-Chibougamau area, Quebec. Roy. Soc. Canada, Trans., 30, sec. IV: 9–12.

63. MILLER, W. G. (1912). Reports on the District of Patricia (a compilation). Ont. Bur. Mines, Ann. Rept. 21, pt. 2.

64. MOOREHOUSE, W. W. (1940). Geology of the Eagle Lake area. Ont. Dept. Mines, 48, pt. 4: 17–18.

65. MULLIGAN, ROBERT (1955). Split Lake, Manitoba. Geol. Surv., Can., Map 10-1956.

66. NIKIFOROFF, C. C. (1952). Origin of microrelief in the Lake Agassiz basin. J. Geol., 60: 99–103.

67. PAULSON, Q. F. (1951). Ground water in the Neche area, Pembina County, North Dakota. North Dakota Geol. Surv. and U. S. Geol. Surv., North Dakota ground water studies, no. 16.

68. ——— (1953). Ground water in the Fairmount area, Richland County, North Dakota and adjacent areas in Minnesota. U. S. Geol. Surv., North Dakota Water Conserv. Comm., and North Dakota Geol. Surv., North Dakota ground water studies, no. 22.

69. PETERSON, R., IVERSON, N. L., and RIVARD, P. J. (1958). Studies of several dam failures on clay foundations. Proceed., Fourth Internat. Conf. Soil Mechanics and Foundation Engineering, London, August 1957, 348–52.

70. QUINN, H. A. (1954). Nelson House, Manitoba. Geol. Surv., Can., Prelim. Map. 54–13.

71. RIDDELL, W. F. (1950). Foundation conditions in Winnipeg and immediate vicinity. Nat. Res. Council of Canada, Assoc. Comm. on Soil and Snow Mechanics, Tech. Mem. 17, 3–9.

72. RIES, HEINRICK, and KEELE, JOSEPH (1912). Preliminary report on the clay and shale deposits of the western provinces. Geol. Surv., Can., Mem. 24-E, 13–30.

73. RITTENHOUSE, G. (1934). A laboratory study of an unusual series of varved clays from northern Ontario. Amer. J. Sci., *228*: 110–20.

74. ROMINGER, J. F., and RUTLEDGE, P. C. (1952). Use of soil mechanics data in correlation and interpretation of Lake Agassiz sediments. J. Geol., *60*: 160–80.

75. RUBIN, MEYER, and ALEXANDER, CORRINE (1958). U.S. Geol. Surv. radiocarbon dates IV. Science, *127*: 1476–87.

76. ———, and ——— (1960). U.S. Geol. Surv. radiocarbon dates V; Amer. J. Sci., Radiocarbon Supplement, 2: 129–85.

77. RUSSELL, L. J. (1958). Geological geomorphology. Geol. Soc. Amer. Bull. *69*: 1–22.

78. SATTERLY, JACK (1937). Glacial lakes Ponask and Sachigo, District of Kenora (Patricia Portion), Ontario. J. Geol., *45*: 790 6.

79. ——— (1938a). Geology of the Stull Lake area. Ont. Dept. Mines, *46*, pt. 4, 1937: 22–6.

80. ——— (1938b). Geology of the Sandy Lake area. Ont. Dept. Mines, *47*, pt. 7: 1–43.

81. ——— (1940). Pleistocene glaciation in the Windigo-North Caribu Lakes area, Kenora District, Ontario. Roy. Can. Inst., Trans., *23*, no. 449, pt. 1.

82. ——— (1943). Geology of the Dryden-Wabigoon area. Ont. Dept. Mines, *50*, pt. 2, 1941: 43–5.

83. ——— (1960). Geology of the Dyment area. Ont. Dept. Mines, *69*, pt. 6: 21–3.

84. SILL, J. (1953). Corrosion and erosion by Winnipeg soils. Nat. Res. Council of Canada, Assoc. Comm. on Soil and Snow Mechanics, Proceed. Sixth Can. Soil Mechanics Conference, Winnipeg, Dec. 1952, Tech. Mem. 27, 21–8.

85. TERASMAE, JAAN (1958). Contributions to Canadian palynology. Geol. Surv., Can., Bull. 46, pt. 2: 13–28.

86. THIEL, G. A. (1947). The geology and underground waters of northeastern Minnesota. Minn. Geol. Surv., Bull. 32.

87. TYRRELL, J. B. (1892). North-western Manitoba with portions of the districts of Assiniboia and Saskatchewan. Geol. and Nat. Hist. Surv., Can., Rept. Progress 1890–91, 5, pt. E.

88. ——— (1896). The genesis of Lake Agassiz. J. Geol., *4*: 811–15.

89. ——— (1913). Hudson Bay exploring expedition, 1912. Ont. Bur. Mines, *22*, pt. 1: 204.

90. ———, and DOWLING, D. B. (1900). Report on the east shore of Lake Winnipeg and adjacent parts of Manitoba and Keewatin. Geol. Surv., Can., Ann. Rept., vol. XI, pt. G (1901).

91. UPHAM, WARREN (1890). Glacial Lake Agassiz in Manitoba. Geol. and Nat. Hist. Surv., Can., Ann. Rept., 4, 1888–89, pt. E.

92. ——— (1894). Preliminary report of field work during 1893 in northeastern Minnesota, chiefly relating to the glacial drift. Minn. Geol. and Nat. Hist. Surv., Ann. Rept. 22, 51.

93. ——— (1895). The Glacial Lake Agassiz. U.S. Geol. Surv., Mon. 25.

94. ——— (1910). Birds Hill, an esker near Winnipeg, Manitoba. Geol. Soc. Amer. Bull., *21*: 407–32.

95. WALLACE, R. C. (1917). The corrosive action of certain brines in Manitoba. J. Geol., *25*: 459–66.

96. ——— (1927). Varve materials and banded rocks. Roy. Soc. Canada, Trans., ser. 3, *21*, sec. IV: 109–18.

97. ——— , and McCartney, G. C. (1928). Heavy minerals in sand horizons in Manitoba and Eastern Saskatchewan. Roy. Soc. Canada, Trans., ser. 3, 22, sec. IV: 199–214.

98. ———, and Maynard, J. E. (1924). The clays of Lake Agassiz basin. Roy. Soc. Canada, Trans., 18, sec. IV: 9–30.

99. Wilson, A. G. W. (1909). Report on a traverse through the southern part of the North-west territories from Lac Seul to Cat Lake, in 1902. Geol. Surv., Can., pub. 1006; also in Ont. Bur. Mines, 21, pt. 2: 49–58 (1912).

100. Wilson, J. T. (chairman) et al. (1958). Glacial map of Canada. Geol. Assoc. Canada.

101. Wright, H. E., Jr. (1956). Sequence of glaciation in eastern Minnesota. Geol. Soc. Amer. Guidebook Series, Field Trip No. 3, Eastern Minnesota, 1–24.

102. Wright, J. F. (1925). Oxford and Knee Lakes area, northern Manitoba. Geol. Surv., Can., Summ. Rept. 1925, pt. B, 16–26.

103. ——— (1927). Island Lake area, Manitoba. Geol. Surv., Can., Summ. Rept. 1927, pt. B, 54–80.

104. Zoltai, S. C. (1960). Glacial history of part of northwestern Ontario. Manuscript in preparation for publication.

The following references deal with the history of Lake Agassiz but do not pertain directly to the soils and are not mentioned in the body of the text:

105. Leighton, M. M., and Wright, H. E., Jr. (1957). Radiocarbon dates of Mankato drift in Minnesota. Science, 125: 1037–9.

106. Leith, E. I. (1949). Fossil elephants of Manitoba. Can. Field-Naturalist, 63, 135–7.

107. Löve, Doris (1959). The postglacial development of the flora of Manitoba: a discussion. Can. J. Botany, 37: 547–85.

108. Mozley, Alan (1934). Post-glacial fossil mollusca in western Canada. Geol. Mag., 71: 370–82.

109. Nikiforoff, C. C. (1947). The life history of Lake Agassiz: alternative interpretation. Amer. Jour. Sci., 245: 205–39.

110. Rudd, V. E. (1951). Geographical affinities of the flora of North Dakota. Amer. Midland Naturalist, 45: 722–39.

111. Rosendahl, C. O. (1948). A contribution to the knowledge of the Pleistocene flora of Minnesota. Ecology, 29: 284–315.

112. Wright, H. E., Jr. (1955). Valders drift in Minnesota. J. Geol., 63, 403–11.

113. ———, and Rubin, Meyer (1956). Radiocarbon dates of Mankato drift in Minnesota. Science, 124: 625–6.

# TILLS OF SOUTHERN ONTARIO

## Aleksis Dreimanis

TILL IS NON-CONSOLIDATED GLACIAL DEPOSIT. It consists of clastic particles, ranging from fine clay to large boulders, and contains all those rocks and other materials that have become incorporated in glacial ice during its motion. Thus the composition of till reveals the bedrock and non-consolidated sediments which have been overridden by glaciers on the way to the place of deposition.

This study deals with only that section of Southern Ontario underlain by Palaeozoic bedrock which is south of an approximate line from the base of the Bruce Peninsula to Thousand Islands. In this region, one to at least five till layers have been encountered, and their total thickness exceeds 300 feet in some bedrock depressions and interlobate areas (Dreimanis, 1953, Sanford, 1953, Watt, 1957). On some bedrock highlands, however, for example along the Niagara escarpment, the till veneer is only a few feet thick. Except in the Toronto area, organic remains of warm temperate climate and deep weathering profiles are absent between tills. This negative evidence suggests that most tills of Southern Ontario were probably deposited during the last or Wisconsin ice age (see Table I, also Dreimanis, 1958, and Dreimanis and Terasmae, 1958).

These tills differ lithologically and texturally not only from one area to another, but even in single sections. Such a variety of tills may be explained by glacial readvances from different directions, since the Pleistocene ice sheet consisted of competing flows. These created lobal protrusions along the ice margin. The glacial lobes followed major bedrock depressions during the last glacial retreat—the Ontario-Erie; the Huron; the Georgian Bay; the Nottawasaga Bay; and the Simcoe-Kawartha Lakes lobe, the area of numerous smaller lake troughs. Similar lobes probably existed also during the earlier oscillations of the Wisconsin ice sheet. Provenance studies of heavy minerals in tills (Dreimanis et al., 1957) have shown that each lobe can be traced back towards the interior of the ice sheet, along flow lines connecting certain associations of heavy detritals in till with their probable bedrock source areas on the Canadian Shield. Each lobe and the corresponding glacial flow inside the ice sheet was influenced by the activity and shifting of centres of glacial outflow. Some of the changes of regional glacial movements during the last ice age have been deciphered by lithologic and fabric studies of tills, crossing striae and superimposed geomorphologic features (Chapman and Putnam, 1951, Dreimanis et al., 1957, Dreimanis and Terasmae, 1958, Dreimanis, 1958, Dell, 1959).

## TABLE I

### TENTATIVE STRATIGRAPHIC SEQUENCE AND CORRELATION OF SELECTED TILLS AND END MORAINES IN THE LAKE ERIE–LAKE ONTARIO LOBAL AREA, WITH THE YOUNGEST TILLS AT THE TOP

(41) Cobourg area upper till  (42) Kingston area upper till

(39) Halton till = (40) Toronto area upper till (Leaside till)

Interstadial, represented by varved clays at Toronto

(36) Galt moraine (36a) }
(35) Paris moraine (35a) } = (38) Wentworth till
(34) Tillsonburg moraine
(33) St. Thomas and Norwich moraine } Port Stanley drift
(32) Westminster moraine = ? = (28) Blenheim moraine
(31) Ingersoll moraine = ? = (27) Charing Cross moraine = ? = (29) Glencoe area till

Lake Erie interstadial in the Lake Erie basin

(22) Catfish Creek till{(22c) = (23) Dorchester mor.} = ? = (24) Dresden lower till = ? = (25) Amherstburg lower till
{(22b)

Plum Point interstadial, ending approximately 25,000 years B.P. in the Lake Erie area

(21) Southwold till

Port Talbot interstadial, ending approximately 44,000 years B.P. in the Lake Erie area

(20) Dunwich till = ? = (44) Toronto lower Wisconsin till

Scarborough beds at Toronto
Sangamon interglacial deposits (Don beds) at Toronto

(45) Illinoian till (York till) at Toronto

(43) Toronto-Cobourg area: middle Wisconsin till

References on terminology: Chapman and Putnam (1951), Dreimanis and Terasmae (1958), Karrow (1959), Dreimanis (1960), Terasmae (1960). Numbers in parentheses refer to Table III.

The principal purpose of the current investigation was: (1) to find out how the composition and texture of tills has been influenced by bedrock and non-consolidated deposits overridden by the ice sheet; (2) to utilize lithologic and textural differences of tills for identifying different lobal areas and different till layers; and (3) to apply results of the above studies in deciphering the Pleistocene history of Southern Ontario.

The present paper is a preliminary report, dealing principally with the first of the above three tasks. This study was begun in 1949, and it has been supported by grants from the Research Council of Ontario and the Geological Survey of Canada. Laboratory and field investigations still continue.

The following properties of tills have been investigated by the author:

1. *granulometric composition*, by using Bouyoucos' hydrometer method (Krumbein and Pettijohn, 1938, p. 173) and assuming 5 $\mu$ and 50 $\mu$ as boundaries between clay, silt and sand;

2. *carbonate content* of till matrix (the −200 mesh or −75 $\mu$ grade), by using the Chittick gasometric apparatus; a new procedure (Dreimanis, (1961) permits also the differentiation of calcite from dolomite[1];

3. *lithologic composition* of the *sand* and the *pebble* grades of till, including also heavy mineral investigations;

4. *till fabric*, as an aid in determination of local glacial movements.

Most results of analyses of types 1, 2, and 3 are summarized in Tables II and III. They are grouped (a) according to natural geomorphologic units (end moraines and the adjoining ground moraine areas behind them: for location of the moraines see Chapman and Putnam, 1951, fig. 11, except for the Dorchester moraine at Dorchester, Ontario); (b) lithologically and texturally distinct till units (such as the Halton and Wentworth tills, see Karrow, 1959); and (c) areas of sampling (for example, the Dresden area). Samples have been taken from surface tills, except for the Catfish Creek, Dunwich, Southwold (see Dreimanis, 1960) and the Toronto Illinoian tills, and those tills which are specified as the "middle" or the "lower" tills. A tentative stratigraphic correlation of tills from the Erie and Ontario lobal area is given in Table I. The author regrets that he could not make direct comparison between his granulometric analyses and those published in Reports of the Ontario Soil Survey, because of differences in the clay/silt boundary.

### Influence of Precambrian Bedrock

All the glacial lobes entered Southern Ontario from the Canadian Shield. As the Precambrian rock fragments, and most of their constituent minerals, are resistant to abrasion, the tills of Precambrian lithology are coarse-grained, sandy and gravelly. An average of three typical till samples from the area

---

[1]Ratios of calcite to dolomite, published by Dreimanis, 1958, and Dreimanis and Terasmae, 1958, were incorrect because of a systematic error in determination of the amount of dolomite; the corrected figures are in Tables II and III of this report (see till groups 20–22, 33, 34, 40, 43, 44, 45).

## TABLE II

AVERAGE RESULTS OF TILL INVESTIGATIONS: LAKE HURON LOBE, GEORGIAN BAY LOBE, NOTTAWASAGA BAY LOBE, SIMCOE AND KAWARTHA LAKES LOBE

| Till area or layer | Granulometric Composition | | | | Carbonates in −200 mesh | | | | | Pebbles (5–25 mm) | | | | | |
|---|---|---|---|---|---|---|---|---|---|---|---|---|---|---|---|
| | Samples anal. | Sand and gran. % | Silt % | Clay % | Samples anal. | Calcite % | Dolom. % | Total carbon. % | calc. dolom. | Samples anal. | Limest. % | Dolom. % | Total carbon. % | Shale % | Ign. and metam. % |
| **A. Lake Huron lobe** | | | | | | | | | | | | | | | |
| 1. Lower till between Forest and Goderich | | | | | | | | | | | | | | | |
| (a) rich in dolomite | 4 | 38 | 31 | 30 | 8 | 22 | 26 | 47 | .8 | 1 | 73 | 16 | 89 | 1 | 6 |
| | | | | | | | | | | 1 | 51 | 13 | 64 | 25 | 6 |
| (b) poor in dolomite | 5 | 42 | 26 | 32 | 5 | 24 | 10 | 34 | 2.4 | 1 | 65 | 4 | 69 | 11 | 6 |
| 2. Mitchell moraine | 9 | 28 | 40 | 32 | 16¹ | 29 | 17 | 46 | 1.7 | 4 | 79 | 15 | 94 | 0 | 5 |
| 3. Lucan moraine | 7 | 24 | 34 | 42 | 20¹ | 30 | 13 | 43 | 2.3 | | | | | | |
| 4. Seaforth moraine | 9 | 14 | 40 | 46 | 24¹ | 24 | 10 | 33 | 2.3 | | | | | | |
| 5. Wyoming moraine | | | | | | | | | | | | | | | |
| (a) N. of Kincardine | 9 | 20 | 34 | 46 | 14 | 16 | 25 | 41 | .7 | 10 | 38 | 45 | 83 | 1 | 11 |
| (b) Lucknow to Grand Bend | 15 | 22 | 34 | 44 | 19 | 22 | 24 | 46 | .9 | 7 | 38 | 32 | 70 | 4 | 13 |
| (c) Grand Bend to Sarnia | 25 | 22 | 32 | 46 | 42¹ | 16 | 19 | 35 | .9 | 2 | 51 | 16 | 67 | 15 | 10 |
| 6. Dresden area: upper till | 2 | 48 | 43 | 9 | 6 | 9 | 12 | 21 | .8 | | | | | | |
| **B. Georgian Bay lobe** | | | | | | | | | | | | | | | |
| 7. Pre-Milverton mor. tills: | | | | | | | | | | | | | | | |
| (a) St. Marys—Stratford area | 6 | 39 | 33 | 28 | 9 | 21 | 21 | 42 | 1.0 | | | | | | |
| (b) Palmerston—Arthur area | 1 | 15 | 35 | 50 | 5 | 15 | 35 | 50 | .4 | | | | | | |
| | 1 | 37 | 39 | 24 | 11 | 20 | 25 | 45 | .8 | | | | | | |
| 8. Milverton mor. | 3 | 16 | 39 | 45 | | | | | | | | | | | |
| 9. Singhampton mor., W. half | 7 | 44 | 32 | 24 | 7 | 18 | 25 | 44 | .6 | 4 | 51 | 35 | 86 | 2 | 10 |
| 10. Gibraltar mor., W. half | 13 | 58 | 28 | 14 | 12 | 15 | 28 | 43 | .6 | 4 | 36 | 48 | 84 | 3 | 10 |

TABLE II (cont'd)

| Till area or layer | Granulometric Composition | | | | Carbonates in −200 mesh | | | | | Pebbles (5–25 mm) | | | | | |
|---|---|---|---|---|---|---|---|---|---|---|---|---|---|---|---|
| | Samples anal. | Sand and gran. % | Silt % | Clay % | Samples anal. | Calcite % | Dolom. % | Total carbon. % | calc. dolom. % | Samples anal. | Limest. % | Dolom. % | Total carbon. % | Shale % | Ign. and metam. % |
| C. *Nottawasaga Bay lobe* | | | | | | | | | | | | | | | |
| 11. Beaver Valley and Nottawasaga basin: | | | | | | | | | | | | | | | |
| (a) sandy till | | | | | 6 | 25 | 9 | 34 | 2.6 | | | | | | |
| (b) clayey till | | | | | 1 | 50 | 3 | 53 | 20 | | | | | | |
| D. *Simcoe and Kawartha Lakes lobe* | | | | | | | | | | | | | | | |
| 12. Simcoe till plain W. of Lake Simcoe | 48[1] | 67 | 23 | 10 | 12 | 26 | 6 | 32 | 4.5 | | | | | | |
| 13. Area W. and E. of Lake Couchiching | 14[2] | 62 | 26 | 12 | | | | | | | | | | | |
| 14. Newmarket area: till older than Lake Schomberg I | 5 | 66 | 21 | 12 | 9 | 27 | 6 | 33 | 4.5 | | | | | | |
| 15. Newmarket area: till younger than Lake Schomberg I | 5 | 40 | 38 | 22 | 7 | 39 | 5 | 44 | 8 | | | | | | |
| 16. Oak Ridges interlob. mor. at Vandorf | | | | | 11 | 37 | 2 | 39 | 20 | 3[3] | | | 85 | 2 | 13 |
| 17. Interlob. mor. and drumlins N. of it between Uxbridge and Belleville | 12[4] | 58 | 23 | 19 | 13 | 44 | 2 | 46 | 20 | 18[5] | | | 90 | + | 9 |
| 18. Dummer mor. | 7[4] | 84 | 11 | 5 | 3 | 15 | 2 | 17 | 7.5 | 10[6] | | | 90 | 2 | 8 |
| 19. Brampton area: lower till | 2 | 52 | 30 | 18 | 3 | 16 | 5 | 21 | 3.2 | | | | | | |

[1]Fraleigh (1954), recalculated. and author.
[2]Deane (1950). p. 17–21.
[3]Henderson (1950).
[4]Gravenor (1957). p. 21. and author.
[5]Henderson (1950). and author.
[6]Henderson (1950). Gravenor (1957). p. 25. and author.

TABLE III

AVERAGE RESULTS OF TILL INVESTIGATIONS: LAKE ERIE AND ONTARIO LOBE

| Till area or layer | Granulometric Composition | | | | Carbonates in −200 mesh | | | | | Pebbles (5–25 mm) | | | | | |
|---|---|---|---|---|---|---|---|---|---|---|---|---|---|---|---|
| | Samples anal. | Sand and gran. % | Silt % | Clay % | Samples anal. | Calcite % | Dolom. % | Total carbon. % | calc./dolom. | Samples anal. | Limest. % | Dolom. % | Total carbon. % | Shale % | Ign. and metam. % |
| 20. Dunwich till | 2 | 62 | 22 | 16 | 5 | 10 | 28 | 38 | .4 | 2 | 33 | 35 | 68 | + | 21 |
| 21. Southwold till | 2 | 47 / 78 | 21 / 14 | 32 / 8 | 6 | 16 | 19 | 35 | .8 | 4 | 42 | 24 | 68 | + | 22 |
| 22. Catfish Creek till | | | | | | | | | | | | | | | |
| (a) base | 2 | 35 | 29 | 36 | 4 | 17 | 10 | 27 | 1.7 | 14[1] | 41 | 38 | 79 | + | 20 |
| (b) lower unit | 20 | 60 | 21 | 19 | 37 | 16 | 21 | 37 | .7 | 10[1] | 51 | 37 | 88 | 0 | 9 |
| (c) upper unit | 17 | 50 | 24 | 26 | 26 | 18 | 17 | 35 | 1.1 | 3[1] | 51 | 24 | 75 | 4 | 8 |
| 23. Dorchester mor. | 8[2] | 43 | 35 | 22 | 16[2] | 18 | 16 | 34 | 1.1 | | | | | | |
| 24. Dresden area: lower till | 2 | 24 | 43 | 33 | 4 | 19 | 15 | 34 | 1.3 | 2 | 15 | 8 | 23 | 73 | 4 |
| 25. Amherstburg area: lower till | 4 | 44 | 29 | 27 | 8 | 18 | 11 | 29 | 1.7 | 1 | 42 | 17 | 59 | 20 | 18 |
| 26. Amherstburg-Leamington area: upper till | 2 | 56 | 23 | 21 | 3 | 9 | 19 | 28 | .5 | | | | | | |
| 27. Charing Cross mor. | 7 | 35 | 23 | 42 | 15 | 16 | 8 | 24 | 2.0 | | | | | | |
| 28. Blenheim mor. | 4 | 30 | 25 | 45 | 12 | 20 | 6 | 26 | 3.6 | | | | | | |
| 29. Glencoe area: upper till | 5 | 12 | 32 | 56 | 11 | 20 | 10 | 30 | 2.0 | | | | | | |
| 30. Kitchener area: upper till | 25 | 25 | 33 | 42 | 3 | 21 | 13 | 34 | 1.8 | 1 | 48 | 25 | 73 | 15 | 6 |
| 31. Ingersoll mor. | 9 | 20 | 29 | 51 | 21 | 25 | 10 | 35 | 2.5 | 10[3] | 49 | 26 | 75 | 2 | 12 |
| 32. Westminster mor. | | | | | 15 | 25 | 8 | 33 | 3.2 | 2[3] | 62 | 17 | 79 | 1 | 18 |
| 33. St. Thomas and Norwich mor. | 19 | 17 | 37 | 47 | 48 | 26 | 8 | 32 | 3.0 | | | | | | |
| | 27 | 15 | 35 | 50 | 44 | 23 | 9 | 31 | 2.6 | | | | | | |
| 34. Tillsonburg mor. | 5 | 65 | 25 | 10 | | | | | | 2 | 26 | 6 | 32 | 58 | 5 |
| 35. Paris moraine: | | | | | | | | | | | | | | | |
| (a) N. of Vanessa | 3 | 19 | 38 | 43 | 5 | 14 | 19 | 34 | .7 | 2 | 17 | 68 | 85 | 0 | 8 |
| (b) S. of Vanessa | | | | | 5 | 22 | 10 | 32 | 2.2 | | | | | | |

TABLE III (Cont'd)

| Till area or layer | Granulometric Composition | | | | Carbonates in −200 mesh | | | | | Pebbles (5–25 mm) | | | | | |
|---|---|---|---|---|---|---|---|---|---|---|---|---|---|---|---|
| | Samples anal. | Sand and gran. % | Silt % | Clay % | Samples anal. | Calcite % | Dolom. % | Total carbon. % | calc. dolom. % | Samples anal. | Limest. % | Dolom. % | Total carbon. % | Shale % | Ign. and metam. % |
| 36. Galt moraine: | | | | | | | | | | | | | | | |
|   (a) N. of Caledonia | 4 | 67 | 27 | 6 | 6 | 16 | 25 | 41 | .6 | 3 | 21 | 65 | 86 | 0 | 8 |
|   (b) S. of Caledonia | 4 | 5 | 26 | 69 | 3 | 23 | 14 | 37 | 1.6 | | | | | | |
| 37. Port Dover area: | | | | | | | | | | | | | | | |
|   lower till | 1 | 58 | 26 | 16 | 2 | 19 | 7 | 26 | 3 | | | | | | |
| | 1 | 34 | 26 | 40 | | | | | | | | | | | |
| 38. Wentworth till | 13[4] | 52 | 32 | 16 | | | | | | | | | | | |
| 39. Halton till | | | | | | | | | | | | | | | |
|   (a) red facies | 13[4] | 32 | 33 | 35 | 10 | 16 | 5 | 21 | 3.2 | 3 | 6 | 19 | 25 | 50 | 4 |
|   (b) gray Dundas facies | 9[4] | 12 | 37 | 51 | 14 | 16 | 5 | 21 | 3.7 | 9[4] | | | 40 | 55 | 4 |
| 40. Toronto-Oshawa area: | | | | | | | | | | | | | | | |
|   upper or Leaside till | 11 | 46 | 25 | 29 | 24 | 30 | 4 | 34 | 7 | 8 | 39 | 1 | 40 | 59 | 4 |
| | 11[5] | 35 | 33 | 30 | | | | | | | | | | | |
| 41. Cobourg area: | | | | | | | | | | | | | | | |
|   upper till | 5 | 64 | 23 | 13 | 9 | 38 | 4 | 42 | 10 | 3[4] | | | 88 | 0 | 12 |
| 42. Kingston area: | | | | | | | | | | | | | | | |
|   upper till | 5 | 62 | 21 | 17 | 7 | 32 | 14 | 46 | 2.2 | 3 | 53 | 10 | 63 | 14 | 11 |
| 43. Toronto–Cobourg area: middle till | 12 | 50 | 30 | 20 | 17 | 26 | 6 | 32 | 4 | 8 | 74 | 2 | 76 | 11 | 11 |
| 44. Toronto area: lower Wisc. till | 4 | 39 | 32 | 29 | 9 | 13 | 8 | 21 | 1.5 | 2 | 81 | 2 | 83 | 7 | 10 |
| 45. Toronto: Illinoian till | 37 | 49 | 21 | 30 | 1 | 13 | 7 | 20 | 2 | 2[a] | 25 | 0 | 25 | 75 | + |

[1]Day (1960), and author.
[2]L. J. Clark, unpublished, and author.
[3]R. M. Vick (1956), and author.
[4]P. F. Karrow, unpublished.
[5]Watt (1957), p. 15–16.
[6]Henderson (1950), and author.
[7]Deane (1950), p. 17–21, Watt (1957), and author.
[8]Henderson (1950), and author.

north of Georgian Bay, for example, contain 71 per cent sand and granules and only 7 per cent clay (see Figure 2). Most of the Precambrian drift materials are non-calcareous, though carbonates have been incorporated from Grenville marbles, producing in a few places slightly calcareous tills (with 1 to 3 per cent carbonates).

Gravenor (1951, p. 69; 1957, p. 24) has estimated from the abundance of heavy minerals in tills and in Precambrian rocks that the tills (nineteen samples) of Southern Ontario contain 7.5—52.5 per cent Precambrian material. The highest percentage (40 to 52 per cent) was found at a distance of fifteen to forty miles from the Precambrian-Palaeozoic boundary (see also Dell, 1959). Most of Southern Ontario tills contain probably less than 15 per cent of Precambrian material.

In tills examined by the author (Dreimanis, 1958, Dreimanis and Terasmae, 1958, and unpublished analyses), Henderson (1950), Knox (1952), Vick (1956) and Day (1960), the pebble grade usually contains 5 to 20 per cent Precambrian rock fragments, and in most cases 10 to 15 per cent. The sand and coarse silt grades (Gravenor, 1951, Dreimanis and Reavely, 1953, Dreimanis, 1958, Dell, 1959) are richer in Precambrian minerals, but their abundance tapers off considerably in the clay grade.

### Influence of Carbonate Bedrock

Calcareous rocks, limestone and dolostone, dominate among the Palaeozoic bedrock of Southern Ontario (see Figure 1). Most of the Palaeozoic rocks, except for the southwestern tip of Ontario, dip slightly west or southwest, thus forming steps and escarpments of various dimensions, facing the advancing glaciers. Such a structure aids glacial quarrying. Carbonate content, particularly in the coarse grades of till, therefore becomes dominant at a distance of a few miles from the Precambrian-Palaeozoic boundary (see Figure 1).

Limestone and dolostone fragments become crushed and abraded during the glacial transport but, being relatively resistant, they can be carried in the coarse grades of till for considerable distances. Tills rich in carbonate fragments, are therefore relatively coarse-textured, sandy and gravelly as a rule (see Figure 2, also Deane, 1950, p. 23, and Gravenor, 1957, p. 22). Carbonates predominate (usually 70 to 90 per cent) in the pebble grade through most of Southern Ontario, as carbonate bedrock belts were traversed repeatedly by the glacial lobes. Only in areas of shale bedrock, and immediately down the direction of glacial flow from them, may the percentage of carbonate pebbles decrease below 50 (see Figure 1).

While examining lithologic composition of separate particle sizes of tills, two carbonate maxima were noticed in all till samples (see Figure 3): (a) in the coarse sand to cobble grades, and (b) in silt and clay grade, with a distinct minimum in the fine sand grade, probably because of more intensive abrasion by quartz and other hard minerals concentrated in the sand grade. The proportions of the two carbonate maxima depend upon distance from the bedrock source. The coarse grade maximum dominates

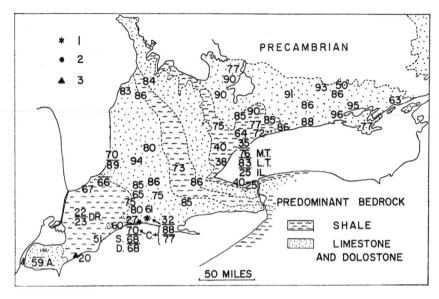

FIGURE 1. Average percentages of carbonate pebbles in till, superimposed upon a simplified bedrock map. Lower tills are represented by figures underneath lines; for instance—A: in the Amherstburg area; DR: in the Dresden area; C: the Catfish Creek till; D: the Dunwich till; S: the Southwold till; M T: the Toronto middle Wisconsin till; L T: the Toronto lower Wisconsin till; I L: the Toronto Illinoian till.

Sample locations for Figure 3: A: 1, B: 2, C: 3.

Bedrock after Sanford, 1957, Goudge, 1938, and Liberty, 1955. Most percentages of calcareous pebbles in the area north of Lake Ontario are after Henderson, 1950.

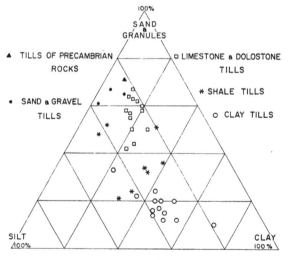

FIGURE 2. Triangular diagram of granulometric composition of five types of till of Southern Ontario, with lines drawn at intervals of 20 per cent. Each dot represents an average of three to fifteen analyses.

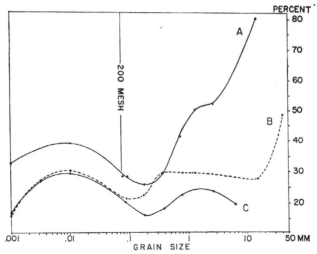

FIGURE 3. Percentage of carbonates versus grain size in three till samples (see Figure 1 for sample locations). A: the Catfish Creek till in limestone bedrock area; B: the upper till; though limestone is the local bedrock according to Sanford, 1957; black shale is present in Lake Erie. C: the upper till in shale bedrock area.

over the fine grade maximum on or near calcareous bedrock areas. At greater distance from the carbonate bedrock source (sample C in Figure 3) the carbonate maximum in the silt and clay grade may become dominant. As the fine grade carbonate maximum is always below the 200 mesh size, the minus 200 mesh grade was chosen for comparison of carbonate content in till matrix.

Till matrix of limestone and dolostone tills contains twice as much carbo-. nate (40 to 50 per cent) as shale tills, except for a narrow Palaeozoic-Precambrian fringe area (see Figure 4). As limestone bedrock areas dominate over dolostone in Southern Ontario (Goudge, 1938), limestone or calcite dominates also in tills, except where areas of extensive dolostone bedrock have been traversed by the glacier (see Figure 5).

*Influence of Shale Bedrock*

Approximately one-fifth of the bedrock sub-outcrop area of Southern Ontario is shale, distributed in several zones, up to fifty miles wide, measured along the direction of known glacial movements (see Figure 1). Most of these shales are friable and with low plasticity, except for the Queenston and the Hamilton shales (Keele, 1924, pp. 14–28). Glaciers could erode them without difficulty. Carbonate content of the Ontario shales is low as a rule: generally less than 15 per cent (Keele, 1924, Caley, 1940, p. 106). Tills rich in shale, therefore usually contain only 20 to 25

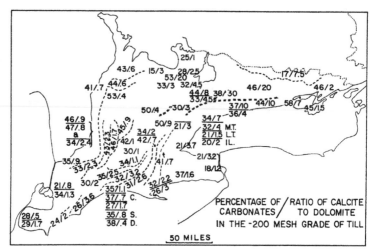

FIGURE 4. Average results of carbonate analyses of the minus 200 mesh grade of tills. See Figure 1 for explanation on lower tills. Selected end-moraines (dashed lines) and interlobate moraines (double dashed lines) after Chapman and Putnam, 1951, Figure 11. Dotted line shows the Palaeozoic-Precambrian boundary.

FIGURE 5. Relation of dolomite rich tills to dolomitic bedrock. Legend—1: dolomitic bedrock; 2: the upper tills; and 3: the lower tills where dolomite dominates over calcite in the minus 200 mesh grade. The only definite boundary between dolomitic tills and tills rich in calcite is the front of the Wyoming moraine; the other boundaries are uncertain. Latest glacial movements (partly after Chapman and Putnam, 1951) are indicated by arrows. Bedrock after Sanford, 1957.

per cent carbonates, or even less in the minus 200 mesh grade (see Figures 4 and 6) except for the Dresden area where the lower till may have been enriched by limestone from the Hamilton formation (Sanford and Brady, 1955, p. 7).

FIGURE 6. Relationship of Dundas shale in tills to the Dundas shale bedrock area at Toronto. Encircled figures show percentage of pebbles of the Dundas shale, siltstone and fine grained sandstone in the upper till (figures with a dot underneath circle are after Henderson, 1950). Figures outside circles give percentage of non-carbonates (derived mostly from shales and Precambrian rocks) in the minus 200 mesh grade. Only two samples (at Lake Ontario; with 90 and 99 per cent shale pebbles) were taken from the contact of till and shale bedrock; others were taken at at least 5 feet, in most cases several tens of feet above bedrock. Principal last glacial movement was from the east towards the west.

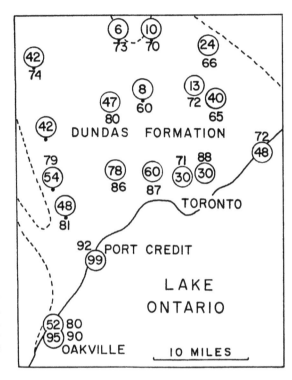

Shale becomes ground to silt and clay-sized particles more rapidly than the carbonate rocks. As a result, shale fragments seldom exceed 60 per cent in the pebble grade of shale rich tills, except for the base of till immediately on shale (see the sample at Port Credit and the lowest sample at Oakville, Figure 6). In the coarse sand grade, abundance of harder varieties of shale may increase up to 70 per cent, as for example in some upper tills along the north shore of Lake Erie (Dreimanis and Reavely, 1953, p. 249; Dell, 1959, p. 191). The grinding of shale fragments by glacial transport to the finest (clay) size of particle requires more than the distance of ten to fifty miles across the shale bedrock. Tills of the shale areas are not therefore predominantly clayey, as often assumed, but have relatively equal average distribution of the clay, silt and sand grades, or even a predominance of either the sand or the silt grades (see Figure 2). Some of the shale formations, as for example the Dundas formation, are rich in clastic particles of the silt and fine sand grade. This may also explain the relative abundance of these two grades in shale tills.

### Influence of Older Pleistocene Deposits

Chapman and Putnam (1951, pp. 10–11) have noted that heavy-textured tills of Ontario are found in areas where the oscillating glacial lobes have overridden stratified clays. Similar overriding and incorporation of sands and gravels has resulted in sandy tills. Principal earlier lake deposits, destroyed during glacial readvances, were in the same areas as the later glacial lakes, resulting in development of clayey tills generally around Lake Huron, Lake Erie, Lake Ontario, and south of Lake Simcoe, with some also in the so-called Ontario Island (northeast of London), which had been occupied by several local temporary lakes.

Incorporation of lacustrine clay results is a typical vertical gradation of clay-rich tills: their base is more finely textured than the top (see Figure 7). This gradation has been found also at the base of some sandy tills, as for example in the lower till at Benmiller (Figure 7). Incorporated fragments of lacustrine deposits, found at the relatively more clayey base of some sandy tills, suggest that the higher clay content is due to admixture of overridden local lake clays. Thus, the distribution of tills with a finer-textured base may assist in mapping the extent of those glacial lake deposits that have been destroyed by subsequent readvances of glaciers.

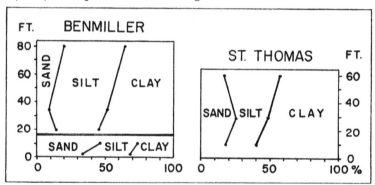

FIGURE 7. Percentage of sand, silt and clay, versus elevation above base of exposure, in two till sections. Benmiller is in the Lake Huron lobe area, St. Thomas in the Lake Erie lobe area.

Coarser-grained outwash, kame, deltaic and shore deposits are more common on the higher ground between the Great Lakes, particularly in interlobate areas. Their overriding by glaciers resulted in the formation of extremely sandy tills (see Figure 2), in some cases with practically no clay grade present, as for example in the Paris moraine near Hespeler (shown to the author by P. F. Karrow). The Paris and Galt moraines serve as good examples of the influence of incorporated non-consolidated glacial deposits. Their northern portions are very sandy, but closer to Lake Erie, where lacustrine deposits were overridden, tills along or behind these moraines become clayey and silty (see Figure 8).

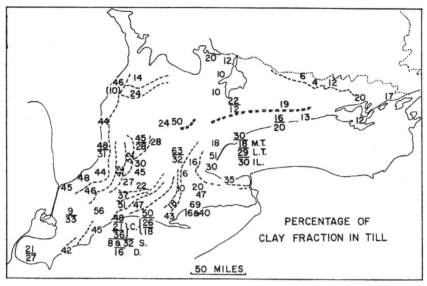

FIGURE 8. Average percentages of the clay fraction in tills. See Figure 1 for explanation on lower tills, and Figure 4 on end-moraines and the Palaeozoic-Precambrian boundary.

Incorporation of Pleistocene deposits of various textures and composition may explain why sandy and clayey tills are found superimposed in some areas, even though the glaciers may have traversed the same bedrock. Thus in the Erie (Dreimanis and Reavely, 1953) and Huron lobal areas, most of the older tills, deposited by the same lobes, are sandy, while the upper tills are clayey. The same relation exists also south of Lake Simcoe, where lacustrine deposits, principally silts, of Lake Schomberg I were overridden by the latest glacial readvance (Dreimanis, 1953). In this area, direct incorporation of any local bedrock material in the uppermost till would be difficult because of the great thickness of the underlying lower till and lacustrine deposits.

Incorporated non-consolidated deposits, if related to a different bedrock source than the till which covers them, influences not only the texture, but also the lithology of the till. For example, the basal portion of the Toronto tills on Scarborough beds is lithologically more similar to the Scarborough beds than to the main body of the same tills (Dreimanis and Terasmae, 1958, p. 123). The same relationship exists also between the carbonate content of the dolomite-rich till of the Wyoming moraine (see Table II) and the underlying lacustrine clays.

*Conclusions and Discussion*

Till is composed of both incorporated older Pleistocene deposits and newly added bedrock fragments. Knowledge of the relative proportions of these components, and of their relation to the respective source areas, will assist in deciphering the history of deposition of each till layer. The scope of this preliminary paper does not permit detailed discussions of each lobal area, its subdivisions and individual till layers, although some textural and lithologic characteristics of forty-five arbitrary units of tills of Southern Ontario are summarized in Tables II and III and on maps 1, 4, 5 and 8. Considerable gaps still exist in our knowledge about some tills, particularly in areas where several glacial lobes were active, as for example to the northeast of London.

Each glacial lobe, if continuously traversing the same type of bedrock, develops certain lithologic characteristics. Tills of the Georgian Bay lobe for example are rich in dolomite (see Table II). Nevertheless, some glacial readvances by the same lobe may produce lithologically completely different tills, particularly because of changes in the location of the centres of glacial outflow. The Huron lobe may be mentioned as an example. There, till of the Wyoming moraine is rich in dolomite, while most of the older tills of the same lobe contain little dolomite (see Table II). Lithologically differing tills may be produced by lobes which have traversed a great variety of bedrock, as for example the Lake Erie–Ontario lobe (see Table III). This variety is increased also by the shifting of the axis of the lobe during the ice age, as suggested by heavy mineral studies (see Dreimanis *et al.*, 1957). The lobes of Lake Erie–Ontario and the Simcoe and Kawartha Lakes seem to have coalesced into one wide glacial flow during some of the Main Wisconsin time, as for example during deposition of the Wentworth till (No. 38 in Table III) and the middle till of the Toronto area (No. 43).

Present knowledge of tills of Southern Ontario suggests a general pattern for their formation. It appears that the major glacial advances, extending southward beyond Southern Ontario, eroded most of the pre-existing non-consolidated Pleistocene sediments first. When deposition began by the same ice sheet, the resulting tills were composed mostly of fresh bedrock fragments. Texturally these tills are medium to coarse grained. Examples are Nos. 1, 14, 19–26, 43–45 (see Tables II and III), except for the very base of some tills such as No. 22a (Table III).

During the general retreat of the Wisconsin ice sheet, its lobes readvanced for shorter distances several times. Tills of these readvances (the most extensive being after the Lake Erie interstadial, see Table I) are rich in reworked local non-consolidated materials. Clays and silts generally became incorporated in the till in areas where lacustrine deposits were overridden, as in Nos. 3 5, 8, 15, 31–34, 35b, 36b, 39b, and some of 40. Meltwater stream and beach deposits, sands and gravels, were incorporated in tills Nos. 17, 23, 35a, 36a, 38.

In each lobal area most of the older ("lower" and "middle") Wisconsin tills were deposited after or during the major glacial advances. They are therefore coarse-textured, while the overlying "upper" tills, representing the latest readvances, are generally clayey or silty in the former lake basins, but with considerable textural varieties in between these lakes, as in the area between London and Lake Simcoe. Such a sequence of impermeable clayey tills, capping relatively permeable coarser-textured tills and stratified drift, is favourable to artesian groundwater conditions. Knowledge of the distribution of drift material of different texture and permeability should assist in groundwater studies. Textural and lithologic studies of tills, if supplemented by other engineering geological and pedological investigations, may be of practical significance also for assistance in engineering and with soil science. Detailed studies of tills of Southern Ontario may therefore have not only theoretical, but also practical application.

### ACKNOWLEDGEMENTS

The author wishes to thank those students of the University of Western Ontario who have participated in analysing tills, and the colleagues who have supplied him with additional till samples and information, particularly C. P. Gravenor, E. P. Henderson, P. F. Karrow. Results of published and unpublished till investigations by other authors have been also included in this report.

### REFERENCES

1. CALEY, J. F. (1940). Palaeozoic geology of the Toronto-Hamilton area, Ontario. Geol. Surv., Can., Mem. 224.
2. CHAPMAN, L. J., and PUTNAM, D. F. (1951). The Physiography of Southern Ontario. Univ. Toronto Press.
3. DAY, C. (1960). Stratigraphy of the lower till in the Catfish Creek area. Dept. Geol., Univ. West. Ont., unpublished senior thesis.
4. DEANE, R. E. (1950). Pleistocene geology of the Lake Simcoe district, Ontario. Geol. Surv., Can., Mem. 256.
5. DELL, C. I. (1959). A study of the mineralogical composition of sand in Southern Ontario. Can. J. Soil Sci., 39: 185–96.
6. Dreimanis, A. (1953). Water. Upper Holland Conservation Report, Ont. Dept. Plan. and Development.
7. ——— (1958). Wisconsin stratigraphy at Port Talbot on the north shore of Lake Erie, Ontario. Ohio J. Sci., 58: 65–84.
8. ——— (1961). Quantitative gasometric determination of calcite and dolomite by using Chittick apparatus. Unpublished manuscript.
9. ———and REAVELY, G. H. (1953). Differentiation of the lower and the upper till along the north shore of Lake Erie. J. Sed. Pet., 23: 238–59.
10. ——— , REAVELY, G. H., COOK, R. J. B., KNOX, K. S., and MORETTI, F. J. (1957). Heavy mineral studies in tills of Ontario and adjacent areas. J. Sed. Pet., 27: 148–61.
11. ——— and TERASMAE, J. (1958). Stratigraphy of Wisconsin glacial deposits of Toronto area, Ontario. Proceed. Geol. Assoc. Can., 10: 119–35.
12. FRALEIGH, R. B. (1954). A study of leaching of the upper till northeast of London, Ontario. Dept. Geol., Univ. West. Ont., unpublished senior thesis.
13. GOUDGE, M. F. (1938). Limestones of Canada, Part IV, Ontario. Dept. of Mines, Ottawa, Rept. No. 781.
14. GRAVENOR, C. P. (1951). Bedrock source of tills in southwestern Ontario. Amer. J. Sci., 249: 66–71.

15. ——— (1957). Surficial geology of the Lindsay-Peterborough area, Ontario, Victoria, Peterborough, Durham, and Northumberland counties, Ontario. Geol. Surv., Can., Mem. *288*.

16. HENDERSON, E. P. (1950). A statistical analysis of the coarse fractions of some Southern Ontario tills. Univ. of Toronto, unpublished Master's thesis.

17. KARROW, P. F. (1959). Pleistocene geology of the Hamilton map-area. Ont. Dept. Mines, Geol. Circular *8*.

18. KEELE, J. (1924). Preliminary report on the clay and shale deposits of Ontario. Geol. Surv., Can., Mem. 142.

19. KNOX, K. S. (1952). The differentiation of the glacial tills along the north shore of Lake Erie. Geol. Dept., Univ. West. Ont., unpublished Master's thesis.

20. KRUMBEIN, W. C., and PETTIJOHN, F. J. (1938). Manual of sedimentary petrology. New York: Appleton-Century Co., Inc.

21. LEGGET, R. F. (1946). A note on Pleistocene deposits of the Sarnia district, Ontario. Trans. Roy. Soc. Can., Ser. 3, *40*, sec. IV: 33–40.

22. SANFORD, B. V. (1953). Preliminary maps. Elgin county and parts of Middlesex county, Ontario, showing drift-thickness and bedrock contours. Geol. Surv., Can., Paper 53–6.

23. ——— compiled by (1957). Geological map of Southwestern Ontario showing oil and natural gas producing areas. Geol. Surv., Can., Map 1062A.

24. ——— and BRADY, W. B. (1955). Palaeozoic geology of the Windsor-Sarnia area, Ontario. Geol. Surv., Can., Mem. 278.

25. TERASMAE, J. (1960). Palynological study of Pleistocene interglacial beds at Toronto, Ontario. Geol. Surv., Can., Bull. *56*: 23–40.

26. VICK, R. M. (1956). The Ingersoll and Westminster end moraines of the London area. Dept. Geol., Univ. West. Ont., unpublished senior thesis.

27. VRIES, H. DE, and DREIMANIS, A. (1960). Finite radiocarbon dates of the Port Talbot interstadial deposits in Southern Ontario. Science, *131*: 1738–9.

28. WATT, A. K. (1957). Pleistocene geology and ground-water resources of the township of North York, York county. Ont. Dept. Mines 64th Ann. Rpt., pt. 7, 1955.

ADDENDUM 1965

LATEST INVESTIGATIONS suggest the following changes in the stratigraphy and correlations of tills in Southern Ontario.

(1) Table I, p. 81: (21) Southwold till is equivalent to the base of the (22) Catfish Creek till, and Port Talbot interstadial ended later than 44,000 years B.P.

(2) Table III, p. 85: (26) Amherstburg-Leamington area upper till was deposited probably by Lake Huron lobe and therefore should be placed in Table II, p. 83, and renumbered as (7) at the end of the group A. *Lake Huron lobe.*

# THE CHAMPLAIN SEA AND ITS SEDIMENTS

## P. F. Karrow

"THE ST. LAWRENCE VALLEY with its north and south extensions along the Ottawa River and Lake Champlain has been studied by geologists for more than half a century and one might expect that its problems would long ago have been solved." This statement was made by A. P. Coleman in his book *The Last Million Years* (1941). A rapidly increasing interest in Pleistocene history during the twenty years since then has resulted in the discovery of many new facts. Many changes in Pleistocene chronology and correlation have been made. Studies in the St. Lawrence valley have played an important part in bringing about these changes.

### PREVIOUS WORK

The importance of the St. Lawrence valley as a source of information on Pleistocene history goes back to 1837 when a paper by Capt. H. W. Bayfield was published in the Transactions of the Geological Society of London. Bayfield had observed elevated marine deposits of clay and sand along the north shore of the St. Lawrence. Shells collected from these deposits by Bayfield were sent to Sir Charles Lyell in England for study. Their similarity to fauna found in clays of Sweden was striking. Following his visit to the St. Lawrence valley in 1842, Lyell concluded that there had been a similar geologic history in both regions during recent time. Lyell (1845) listed a number of species of shells found in the marine formations and described many features of the glacial geology. The occurrence of marine clays containing boulders led him to the conclusion that sea-borne ice had been an active agent of deposition.

In 1855, Sir William E. Logan of the Geological Survey of Canada gave to J. W. Dawson the task of studying the "superficial" deposits of Canada. Logan's *Geology of Canada* (1863) incorporated the results of Dawson's studies up to that time. Dawson referred to Lyell as his "guide and instructor" and, like Lyell, believed that sea-borne ice had been the agent of deposition for the glacial drift of the St. Lawrence valley. He divided the strictly marine deposits into Leda clay (deep-water sediments) and Saxicava sand (shallow-water sediments) based on lithology and the fossils they contained. Perhaps the greatest single contribution was the listing of the extensive fauna of the marine sediments. In 1863 the list contained eighty-three species and by 1893, when his *The Canadian Ice Age* was published, he had extended it to about two hundred species.

By the turn of the century, the theory of continental glaciation was widely held and multiple glaciations were being referred to. Keele noted that there were several kinds of clay deposits in the St. Lawrence valley and that there was no necessary correlation between elevation and type of deposit. His studies of clay deposits (1915, 1924) are still standard references.

A. P. Coleman devoted much attention to glacial and marine features in the Champlain Sea area over a period of forty years. His publications include studies of raised beaches (1901) and what he interpreted as inter-bedded till and marine sediments (1927). *The Last Million Years* (1941) summarized his work on the Champlain Sea.

Reports on the areal distribution of Pleistocene deposits in the Montreal and Ottawa areas were published by the Geological Survey of Canada in 1915 and 1917 respectively. J. Stansfield, describing the Montreal area, recorded an upper marine level of 617 feet above the present sea level. W. A. Johnston, in the Ottawa report, recorded a level of 690 feet near Ottawa.

A regional study by E. Antevs (1925) of the retreat of the last ice-sheet, as shown by correlation of varved clays, included observations on the marine sediment sequences at Ottawa. A second marine invasion, known as the Ottawa Sea, was postulated and its upper limits were placed at 470 feet. Antevs further concluded (in agreement with Johnston, 1916) that a fresh-water stage followed in the Ottawa area. Lake Ottawa, as he called it, cut shorelines whose elevation varied from 260 feet in the northwest to 230 feet in the east near its outlet at Hawkesbury.

Extensive studies of raised shorelines were made by J. W. Goldthwait (1933). He defined the upper marine limit at many localities and worked out the pattern of sea retreat. Prominent beaches at 210 and 110 feet above present sea level indicated to him interrupted uplift of the land. He assigned lower features to the river stage.

A period of renewed interest in the St. Lawrence valley began in 1950 with the definition and description of the St. Narcisse moraine by F. F. Osborne. J. Beland in describing the geology of the Shawinigan Falls area (1953) described an exposure showing till over marine sediments along the axis of the St. Narcisse moraine, correctly inferring a readvance of the ice into the Champlain Sea.

Meanwhile, in 1950, the Geological Survey assigned N. R. Gadd to map the Becancour area in the central part of the St. Lawrence lowlands. Subsequently, the adjacent areas were mapped as well, six areas all told being completed by 1956. Montreal Island was remapped during this period and, soon after, the Ottawa, Cornwall, Renfrew, and Chalk River areas were studied. Concurrently with the systematic mapping programme, special studies on palaeontological subjects were undertaken. Miss F. J. E. Wagner made extensive collections of the abundant fauna of the marine sediments, and J. Terasmae studied interglacial and post-glacial spores and pollen. Some of the results of these investigations have been published (Owen, 1951; Terasmae, 1958, 1959, 1960; Karrow, 1959; Gadd, 1960a, 1960b; Gadd and Karrow, 1960). Others are to appear in the near future or are in preparation. Other significant studies of the last decade are those of Chapman and Putnam (1951), Elson and Elson (1959), Allen and Johns (1960), and MacClintock and Terasmae (1960).

Specific findings of recent studies will be now discussed, grouped into appropriate topics.

### EXTENT OF THE CHAMPLAIN SEA

The maximum extent of the marine inundation is as yet known only in general terms. There are good reasons for this. Much of the area concerned is within the Precambrian Shield where the glacial drift is thin and the bedrock is resistant to wave attack. Material to form constructional shoreline landforms is therefore in short supply, and erosional features are little-developed. Over much of the highlands where the highest water levels should be recorded, there is a dense cover of vegetation which conceals such features as beach ridges. Marine features are best distinguished from freshwater features by the presence of marine fossils and yet these are poorly preserved within the area of the Shield because of the acid rocks and rapid leaching of carbonates. Often only moulds of shells remain and they are easily missed. The irregular topography of the Shield formed a very irregular shoreline with innumerable islands; wave action would be damped by such a coastline. Last, but not least, it is possible that the highest levels of marine water existed when ice still covered part of the St. Lawrence valley and formed the north shore of the sea. The location of the shoreline at any time is a function of the level of the land (subject to isostatic adjustment during and after glaciation) and the level of the sea (eustatic changes in sea level subject to climate and the amount of water frozen in glaciers).

Chapman and Putnam (1951) state that "the first fact to record about the Champlain Sea is that, because of the rocky nature of the slope, it did not have shore features along its western shore." Marine fossils have been found as far west as Brockville, Smiths Falls, Arnprior, and near Pembroke.

The northern shoreline is very irregular, extending into the Laurentian highlands north of Ottawa, Montreal, and Quebec City. The Champlain Sea was open to the Gulf of St. Lawrence on the east but, for convenience, a boundary may be placed between the two at the constriction between highlands to the north and south near Quebec City. East of that boundary, conditions are still transitional to marine environment. The former southern shoreline of the Champlain Sea is at a lower elevation than the northern because of the steep gradient southward down the isobases of isostatic uplift. A long embayment extended down the Lake Champlain depression. West of this, the shoreline headed around Covey Hill at 525 feet and on toward the Lake Ontario basin.

It has been thought that water in the Lake Ontario basin was at one time at the same level as the Champlain Sea and formed a large bay open to the east (e.g. Mather, 1917). A re-examination of the evidence for this should be made. In any case, such a bay was probably nearly, if not completely, fresh and it may be diregarded in this discussion.

### STRATIGRAPHIC POSITION AND CHRONOLOGY

The stratigraphic sequence for the central St. Lawrence lowlands is shown in Figure 1. No evidence of marine episodes earlier than the Champlain Sea has been found during recent studies. The sea apparently entered the valley

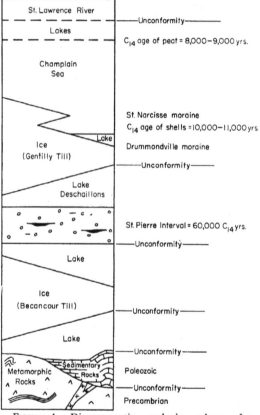

FIGURE 1. Diagrammatic geologic column for central St. Lawrence valley

after partial deglaciation had taken place since, in the southern part of the lowlands, non-marine sediments often lie between the glacial deposits below and overlying marine sediments (Gadd, 1960a, 1960b; MacClintock and

Terasmae, 1960). As the ice retreated northward, its margin stood in the sea. After retreating an unknown distance north of the lowlands, the ice readvanced to form the St. Narcisse moraine, overriding fossiliferous marine sediments (Karrow, 1959). Final retreat of the ice then left the area inundated by the sea.

It has long been known that the Champlain Sea invasion followed the retreat of the last ice-sheet. Extrapolation from the central Great Lakes area indicated what would now be known as a post-Valders age for the Champlain Sea. A recent summary of these views was given by Hough (1958) who suggested 8,000 years before present for the beginning of the Champlain Sea. More recent carbon-14 dating on peat from post-marine bogs now indicates a greater age. Currently, the beginning of the marine episode is estimated at 11,000 to 12,000 years and its end at about 8,000 to 9,000 years ago (Terasmae, 1959, 1960; Gadd, 1960). Major revisions in glacial and Great Lakes history seem to be necessary.

The Champlain Sea came to an end after uplift of the land exceeded the eustatic rise of sea level. An ever-diminishing volume of sea water in the St. Lawrence valley became freshened by rivers flowing into the lowlands. The change from marine to freshwater conditions was no doubt gradational. Multiple marine stages at Ottawa have not been supported by recent field work (N. R. Gadd, personal communication). There is evidence that one or more temporary lakes were left during the draining of the St. Lawrence and Ottawa valleys (Johnston, 1916; Antevs, 1925; Elson and Elson, 1959). Lac St. Pierre, between Sorel and Trois Rivières, Quebec, is a modern example.

## Marine Deposits

The sediments deposited in the Champlain Sea are closely associated with glacial deposits. It is important to remember that the sea bottom was largely a glacial landscape which was flooded as the ice retreated northward. Ice formed the northern shore of the sea for an unknown length of time, depositing its load of sediment in the sea as it retreated.

Probably the most obvious feature in areas underlain by marine deposits is the nearly level plain which forms the surface. The plain is not truly level or flat, but is concave upward. The degree of curvature varies noticeably with the distance between the boundaries of the plain. Where there is a narrow embayment, the plain surface slopes upward around the edges to a marked degree and tends to approach the adjacent hill surfaces tangentially, although this result is not achieved. The curvature is probably a function of initial dips on an irregular surface, the degree of tranquillity of the waters during the history of sedimentation, and compaction of the sediment after deposition. This last factor would vary with the lithology and thickness of the deposit, being less in thin silty or sandy deposits.

Each marine basin near the margin of the sea was partly filled with sediment; the elevation of the surface plain varies somewhat from basin to basin. These variations, however, are usually minor compared to the irregularities of the bottom of the marine deposits. The thickness of the marine deposits varies greatly, mainly because of the conformation of the lower surface (see Figure 2).

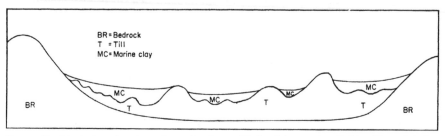

FIGURE 2.    Typical section through overburden and bedrock in area invaded by Champlain Sea

Some typical thicknesses of marine deposits reported in various areas are indicative of the range to be expected. Owen (1951) reports 35 to 80 feet of clay in the deeper basins of the Cornwall-Cardinal area of eastern Ontario. The Hydro-Electric Power Commission of Ontario, in its excavations for the St. Lawrence Power Development, found a maximum depth of about 100 feet but usually less than 50 feet of clay. Johnston (1917) reported a maximum thickness of nearly 200 feet in the Ottawa area and several nearby localities with 100 to 200 feet of clay. On Montreal Island, Stansfield (1915) found relatively thin deposits, the maximum being 79 feet. McGerrigle (1938) reported a maximum of 145 feet in the Lachute area of Quebec. Farther east, in the Becancour area, near Trois Rivières, Gadd reports (1960a) many deposits more than 100 feet thick. Immediately to the north in the Grondines area, the writer has observed exposures of 65 feet of marine sediments and has heard of more than 100 feet of clay being encountered in water wells.

The marine sediments were derived mainly from three sources. In order of decreasing quantitative importance they are: streams, shoreline erosion, and glacial ice. Streams at first flowed directly from glacial ice, later from land uncovered by the retreating glacier, and finally from land uncovered by the retreating shoreline of the sea. Shoreline erosion acted first upon glacial deposits, and later on marine deposits as well, at successively lower elevations as the land rose isostatically. Glacial ice supplied sediment by ablation while its margin stood in the sea.

In considering the nature of the marine deposits, the importance of facies changes must be emphasized. The fine-grained sediments ("Leda clay" of Dawson) constitute the bulk of the marine deposits but vary greatly in character. Proximity to the source of sediment controlled texture and bedding thickness (see Figure 3). As noted above, streams were the most important source of sediment and their effects will be considered first.

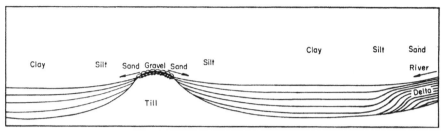

FIGURE 3.   Facies of marine sediments. Arrows indicate movement of sediments from sources.

Where streams entered the sea, deltas were built whose size varied with
the quantity of sediment supplied, the texture of the sediment, and the
strength of currents along the shore near the stream mouth. A large delta
was constructed at the mouth of the St. Maurice River (in the Trois
Rivières area). Its structure shows a similar sequence vertically and laterally.
In general, the sediments grow finer downward and outward toward the
periphery. Bedding thickness increases in a parallel fashion, thin bedding
being associated with interbedded silt and sand layers and thicker, more
massive, beds being formed from clay and silt. Near the top of the sequence,
ripple marks give evidence of relatively shallow water, while contortions in
certain beds may be attributed to underwater slumping. Marine shells are
usually not abundant in deltaic deposits, no doubt because of the freshness of
the water near river mouths. Thin-bedded silt and sand has been found near
other river mouths and may be considered a type of sediment characteristic
of the deltaic environment. The extreme uniformity of bedding thickness
through many feet of sediment is reminiscent of varving as found in glacial
lake deposits. With increasing distance from the river mouth, the sand layers
thin to mere partings, while the fine layers change from silt to clay and
thicken to a foot or more.

Wherever glacial deposits appeared above the surface, they were attacked
by waves and their erosion supplied sediment. The glacial deposits themselves were often reduced to a lag deposit of boulders and gravel, while
sand, silt, and clay were carried into deeper water. Thin-bedded silt and
sand also occur near former shorelines and shoals. Probably because the
salinity of the water was similar to deeper parts of the sea, fossils are sometimes abundant. Stratification in these deposits is usually much less regular
than in deltas.

While the margin of the ice-sheet stood in the sea, coarse sediment was
deposited which had a texture between glacial till and marine clay (Karrow,
1957). The characteristics vary, reflecting varying conditions of sedimentation. No doubt icebergs and shelf ice contributed some sediment. Sometimes
marine deposits were reworked when ice overrode them, as was the case
in the St. Narcisse moraine. The sediments which resulted from this glacio-
marine environment can be described as stoney marine clays or silts, or as

fossiliferous till, depending on which end-member of the series they approached. All variations from one to the other can be found. They seem to occur most commonly near the St. Narcisse moraine, along the edge of the Laurentian upland.

The coarse-grained marine sediments include much of Dawson's "Saxicava sand." They are largely beach and delta deposits of fine to medium sand and, near till outcrops, fine to coarse gravel, subangular to well-rounded. The sands are usually 10 to 20 feet thick although accumulations of 75 feet or more are present in deltas. The major sediment sources were outwash streams flowing into the Champlain Sea from the north so the sediments are merely ground-up Precambrian rocks and are very low in carbonate minerals. This fact, together with the high permeability of the well-sorted marine sands, explains the general lack of fossils in beach deposits in Quebec north of the St. Lawrence River. South of the river, enough Palaeozoic limestone has been incorporated to allow their preservation. North of the river, therefore, identification of sands as marine depends on their location and elevation.

## COMPOSITION OF MARINE CLAYS

The unusual physical properties of the marine clays and silts (in particular, their sensitivity or loss of strength when remoulded) has stimulated much interest in their mineralogy. Several research workers have tackled the problem and yet only partial agreement has been reached.

Related to the problem of mineralogy is proper definition of the grain-size distribution. It is well known that mineral composition varies with different ranges of grain sizes. Some attempts have been made to define these variations in composition for marine clays (Moretto, 1949; Grim, in Terzaghi and Peck, 1948). In general, it was found that the silt portion was largely quartz and calcite and the clay portion (finer than two microns) was mica-illite and montmorillonite. These results were probably obtained by differential thermal analyses.

Other analyses have yielded a variety of results. A sample from the famous St. Thuribe landslide was reported to be 64 per cent silt and 36 per cent clay and to contain mostly "very fine quartz grains with a little mica and possibly a trace of montmorillonite" (Grim, in Peck, Ireland, and Fry, 1951). The high silt content is no doubt caused by the proximity of the site to the St. Narcisse moraine and is not necessarily typical of marine sediments. Beland reported (1956) mostly quartz and feldspar, subordinate mica, and a very small quantity of clay minerals, commonly montmorillonite and kaolinite, at the site of the Nicolet landslide.

Twelve samples from the Grondines map-area (including St. Thuribe) and one from Nicolet were studied by X-ray diffraction of the clay portion, and by mechanical analyses (Karrow, 1957). Although proportions varied somewhat, and carbonates were sometimes present, chlorite, illite, amphibole, quartz, and feldspar were identified in all samples. Illite was dominant in coarser sediment, chlorite in finer. Quartz and feldspar were each probably less than 10 per cent of the clay fraction. Tills and freshwater varved clays in the same area were found to be similar in composition.

Allen and Johns (1960) recently reported on three clay samples from the Champlain Sea deposits and found abundant to moderate hydrous mica, moderate to no chlorite, abundant to minor vermiculite, moderate montmorillonite in one sample, mica-chlorite mixed layers in one sample, abundant quartz, minor to trace of feldspar, and amphibole in two samples. These samples also were run by X-ray diffraction. They concluded that "varved clays deposited in lakes have the same clay-mineral composition as clays containing Quaternary marine fossils and Palaeozoic marine shales in the same area."

Examination of the foregoing results shows that more study is needed adequately to define the composition of the marine deposits.

POST-DEPOSITIONAL HISTORY

Isostatic uplift has elevated the land far above its former depressed position while under the continental glacier. All of the St. Lawrence River valley west of Quebec City is now above sea level. Uplift is probably continuing, but at a much reduced rate. As on any land surface, streams are eroding valleys. Most of the area is still in the youthful stage of the erosion cycle and much undissected upland remains between valleys. Most streams are youthful and many temporary base-levels exist along the stream valleys. Steep gradients, V-shaped valleys, and waterfalls are common. Erosion is occuring quite rapidly in the marine clays. Surface run-off is high because of the impermeability of the deposits, although the clay itself is not easily eroded merely by water running over its surface. Erosion is aided by weathering and desiccation of the banks which causes cracking and collapse of small blocks of clay. Frost action may also contribute to this process.

No doubt the most spectacular type of event which contributes greatly to erosion of the marine clays is the "flow slide." This type of landslide is common with extra-sensitive clays and is well known in the history of the St. Lawrence valley. Although not all landslides in the valley are of the flow type, they are probably the most important areally and volumetrically. They must be considered a major process of valley widening since they have caused the lowering of many square miles of marine plain. They occur where streams are actively eroding their banks, as on the outside curve of a meander. The bank must be sufficiently high to cause instability of the slope, the first step in the development of a flow slide being a slope failure. When this occurs, the clay is remoulded and flows away downstream as a fluid. Slope failure occurs again and again until sufficient material has accumulated in front to bring about stability. During the flow, a pear-shaped depression tends to form. The hummocky surface of the floor of the depression consists of blocks of clay which have not been remoulded, or sand. These hummocks usually represent the weathered upper few feet of the marine sediments. The surface of the depression gradually becomes smoothed out as the hummocks are eroded and weathered down. These flow slide depressions are easily identified on aerial photographs. Although most slides have caused only property damage, some have resulted in the loss of several lives.

## Suggested Research

Although some research problems may be inferred on the basis of more or less negative evidence, it would be well to list here some of the particular problems that the writer considers deserve attention.

1. Search for marine fossils and sediments west of Brockville, Ontario, to indicate the relation of the Champlain Sea to the Lake Ontario basin. Was the sea level high enough (or the land low enough) to permit marine water to enter the Lake Ontario basin?

2. A comprehensive study of the mineralogy of marine clays and silts. This would require closely relating grain-size and composition because they must both be defined for any particular sample. Electron micrography might be of considerable aid.

3. Determination of the pattern of salt concentration, to relate it to initial conditions (marine or freshwater) and weathering (leaching). This could also be related to the base-exchange capacity of the clays and to (2) above.

4. Statistical study of bedding thickness and grain-size distribution.

5. Bottom coring in existing lakes (for example, Lac St. Pierre) to find the sequence of sediments—there may be a record of glacial, freshwater, marine, and final freshwater stages.

6. Regional study of the Champlain Sea area by study of aerial photographs to locate flow slides. As many as possible should be dated by inter-viewing. This has been done for the Grondines area (Karrow 1957, 1959). In many of the slides, wood may be found which can be dated by carbon-14 analysis and the records can thereby be extended into prehistoric time.

7. An isopach map of marine deposits, compiled from well and boring records would be of considerable interest to those in the construction fields.

## ACKNOWLEDGEMENTS

The writer wishes to acknowledge helpful discussions on problems of the Champlain Sea with many persons, but in particular Miss F. J. E. Wagner, N. R. Gadd, and J. Terasmae of the Geological Survey of Canada, and J. A. Elson of McGill University.

## REFERENCES

1. ALLEN, V. T., and JOHNS, W. D. (1960). Clays and clay minerals of New England and Eastern Canada. Geol. Soc. Amer. Bull., 71: 75–86.
2. ANTEVS, E. (1925). Retreat of the last ice-sheet in Eastern Canada. Geol. Surv., Can., Mem. 146.
3. BAYFIELD, H. W. (1837). Notes on the geology of the north coast of the St. Lawrence. Trans. Geol. Soc., London, 2nd ser., 5: 89–103.
4. BELAND, J. (1953). Geology of the Shawinigan map-area, Champlain and St. Maurice Counties, Quebec. Ph.D. thesis, Princeton University.
5. ——— (1956). Nicolet landslide. Proc. Geol. Assoc. Canada, 8, pt. 1: 143–56.
6. CHAPMAN, L. J., and PUTNAM, D. F. (1951). The physiography of southern Ontario. University of Toronto Press.
7. COLEMAN, A. P. (1901). Sea beaches of Eastern Ontario. Tenth Rept. of Ont. Bur. of Mines, 215–27.
8. ——— (1927). Glacial and interglacial periods in eastern Canada. J. Geol., 35: 385–403.
9. ——— (1941). The Last Million Years. University of Toronto Press.
10. DAWSON, J. W. (1893). The Canadian ice age. Montreal: William V. Dawson.
11. ELSON, J. A., and ELSON, J. B. (1959). Phases of the Champlain Sea indicated by littoral molluscs. Geol. Soc. Amer. Bull., 70: 1596, abstract.
12. GADD, N. R. (1957). Geological aspects of Eastern Canadian flow slides. Proc. Tenth Can. Soil Mech. Conf., Nat. Res. Council Tech. Mem. 46, 1956, 2–8.
13. ——— (1960a). Surficial geology of the Bécancour Map-area, Quebec. Geol. Surv., Can., Paper 59–8.
14. ——— (1960b). Surficial geology, Aston, Quebec. Geol. Surv., Can., Map 50–1959.
15. ——— and KARROW, P. F. (1960). Surficial geology, Trois Rivières, Quebec. Geol. Surv., Can., Map 54–1959.
16. GOLDTHWAIT, J. W. (1933). The St. Lawrence Lowland. Unpublished ms of the Geol. Surv., Can.
17. HOUGH, J. L. (1958). Geology of the Great Lakes. University of Illinois Press.
18. JOHNSTON, W. A. (1916). Late Pleistocene oscillations of sea-level in the Ottawa valley. Geol. Surv., Can., Mus. Bull. 24.
19. ——— (1917). Pleistocene and Recent deposits in the vicinity of Ottawa, with a description of the soils. Geol. Surv., Can., Mem. 101.
20. KARROW, P. F. (1957). Pleistocene geology of the Grondines map-area, Quebec. Ph.D. thesis, University of Illinois.
21. ——— (1959). Surficial geology, Grondines, Quebec. Geol. Surv., Can., Map 41–1959.
22. KEELE, J. (1915). Preliminary report on the clay and shale deposits of the Province of Quebec. Geol. Surv., Can., Mem. 64.
23. ——— (1924). Preliminary report on the clay and shale deposits of Ontario. Geol. Surv., Can., Mem. 142.
24. LOGAN, W. E. and others (1863). Report on the geology of Canada. Geol. Surv., Can., Rept. of Progress to 1863.
25. LYELL, C. (1845). Travels in North America in the years 1841–1842, with geological observations in the United States, Canada, and Nova Scotia. New York.
26. MacCLINTOCK, P., and TERASMAE, J. (1960). Glacial history of Covey Hill. J. Geol., 68: 232–41.
27. MATHER, K. F. (1917). The Champlain Sea in the Lake Ontario basin. J. Geol., 25: 542–54.
28. MORETTO, O. (1948). Effect of natural hardening on the unconfined compressive strength of remoulded clays. Proc. International Conf. on Soil Mech. and Foundation Eng., Rotterdam.

29. McGERRIGLE, H. W. (1938). Lachute map-area, Part II: the Lowland area. Que. Bur. of Mines, Ann. Rept. 1936.

30. OSBORNE, F. F. (1950). Ventifacts at Mont Carmel, Quebec. Trans. Roy. Soc. Can., Series 3, *44*, sec. IV: 41–9.

31. OWEN, E. B. (1951). Pleistocene and Recent deposits of the Cornwall-Cardinal area. Geol. Surv., Can., Paper 51–12.

32. PECK, R. B., IRELAND, H. O., and FRY, T. S. (1951). Studies of soil characteristics, the earth flows of St. Thuribe, Quebec. University of Illinois Soil Mechanics, Series No. 1.

33. STANSFIELD, J. (1915). The Pleistocene and Recent deposits of the island of Montreal. Geol. Surv., Can., Mem. 73.

34. TERASMAE, J. (1958). Contributions to Canadian palynology. Geol. Surv., Can., Bull. 46.

35. — – (1959). Notes on the Champlain Sea episode in the St. Lawrence lowlands, Quebec. Science *130*: 334–6.

36. ——— (1960). Contributions to Canadian palynology, No. 2. Geol. Surv., Can., Bull. 56.

37. TERZAGHI, K., and PECK, R. B. (1948). Soil Mechanics in engineering practice. New York: John Wiley and Sons.

# GLACIAL GEOLOGY AND THE SOILS OF NOVA SCOTIA*

## H. L. Cameron

NOVA SCOTIA has been very completely mapped geologically by the Geological Survey of Canada and other agencies, and physiographically by the Topographic Survey of Canada. During the past twelve years the Dominion and Provincial departments of Agriculture have carried on a joint soil survey programme which has now carried out a reconnaissance on the entire province. Over the past ten years the Nova Scotia Research Foundation has carried on a reconnaissance Pleistocene survey which has now covered the entire province. A new 1-inch/4-mile topographic map with 100-foot contours has just been completed by the Topographic Survey.

With all this information available, it is possible to make a good correlation between the bedrock geology, the effects of glaciation on this geology, and the soils which have developed on the glacial material. As might be expected, there is a close connection between soils and bedrock. The glaciation appears to have caused surprisingly little soil transportation with the result that the soils map and bedrock geology map are nearly identical. So close is the relationship that in some cases those working on soils have made suggestions as to age correlations of doubtful areas, which later proved to be correct. In turn geologists have been able to point to some areas where the soils should be identical because of the geology and soil scientists have found that their apparent differences were mainly of particle size caused by local topographic relief. In at least one instance, the soil survey differentiated a facies within a rock group which has only recently been noted by the geologists.

*Bedrock Geology*

The geological map of Nova Scotia presents a complex mosaic which represents rocks ranging in age from Precambrian to Triassic (see Figure 1). These include the common sedimentary rocks, such as shales, mudstones, sandstones and limestones. Metamorphic rocks include slates, quartzites, schists, gneisses, and marbles. Intrusive igneous rocks are represented by large granite masses, smaller diorite and syenite plugs, sills and dikes. Extrusive lavas, both acidic and basic, occur in many places, the most spectacular being the series of Triassic basalt flows which make up North Mountain.

*Physiography*

All these rocks differ in hardness and the amount of folding which they have undergone, both factors being important in determining the resistance

*Published with the permission of the Nova Scotia Research Foundation.

# GENERALIZED GEOLOGICAL MAP OF NOVA SCOTIA

SCALE — MILES

BASED ON COMPILATION BY PROF. N. L. CAMERON
NOVA SCOTIA RESEARCH FOUNDATION

TO ACCOMPANY "GEOLOGY, GLACIAL GEOLOGY AND SOILS OF NOVA SCOTIA"

BY N. L. CAMERON
1960

LEGEND

of the particular rock to erosion. It may be noted that folding is more important than rock hardness in determining resistance to erosion. Examples are the synclines and synclinoria in the Meguma Series, where slate synclines form residual ridges while adjacent quartzite anticlines are as much as two hundred feet lower. The dip of the limbs of a fold does not appear to be significant, since the same conditions hold along the coast of Prince Edward Island where dips are mainly under 15°. Coastal points are underlain by synclines, while bays are anticlinal in structure. Another example is the coastline of the Sydney Coal Field.

The total expression of the varying hardness of the rocks is the physiography of the province. This is clearly shown in the new 1-inch/4-mile topographic map of the province. Briefly, the present surface represents the result of peneplanation, probably in the late Triassic or early Jurassic, and the etching of the peneplane after re-elevation at the close of the Cretaceous. The sensitivity of the topography to even slight changes in rock hardness, or resistance due to folding, is remarkable. The topography has also a great influence on the soils, particularly with regard to the size of the rock fragments present in them. For example, near uplands the lowland soils tend to be stonier, with many of the stones coming from the more resistant upland rocks. This is a local effect however; in general the soils are made up of fragments of the rocks on which they rest, and are stoney, or stone free, depending on the characteristics of the bedrock.

### Soils

Figure 2 is a generalized soils map of Nova Scotia. This map, except for the redraughting, is entirely the work of Dr. Bruce Cann of the Canada Department of Agriculture and the Nova Scotia Soil Survey. The divisions are based on a combination of soil type, physiographic type, and environmental type. The actual divisions are not stated since they are not finalized and also represent information which has not yet been published. This may seem at first glance to be a classification which, although it may be admirably suited to the practical application of soil divisions, would have little value for the correlation of bedrock geology, glaciation, and soils.

Proof of its general soundness is demonstrated when the broad divisions of soil types are compared with the geological divisions (see Figures 1 and 2). When the smaller divisions are compared, the correlation is good. The greatest differences are found where large areas, such as the granite mass of central and western Nova Scotia, are subdivided on the basis of environmental types, for example, well drained, swampy, etc. A fair generalization would be to say that the divisions of soils and bedrock areas are closely related and show little effect from the Continental and local glaciations which have affected the area of Nova Scotia.

### Glaciation

A general surface map of the glacial geology of Nova Scotia has not yet been prepared, but study of the Pleistocene deposits of the province is

# GENERALIZED SOILS MAP OF NOVA SCOTIA

SCALE

COMPILATION BY D.B. CANN, CANADA DEPARTMENT OF AGRICULTURE

CARTOGRAPHY BY NOVA SCOTIA RESEARCH FOUNDATION

**LEVEL TO VERY UNDULATING FINE SANDY LOAMS TO SILTY CLAY LOAMS, STONE FREE**
ACADIA, STEWIACKE, CUMBERLAND

**LEVEL TO GENTLY ROLLING LOAMY SANDS, SANDY LOAMS AND GRAVELLY SANDY LOAMS**
CORNWALLIS, CANNING, HATFIELD, TERREBONNE, BERWICK
HEBERT, CUMBERLAND, PARRSBORO

**UNDULATING TO GENTLY ROLLING SANDY LOAMS - MOSTLY WELL DRAINED**
TORBROOK, PUGWASH, SWANSFORD
TRURO, FURNACE, HANSFORD
WOODVILLE, SOMERSET, PELTON

**UNDULATING TO ROLLING STONY SANDY LOAMS - SOME STEEP SLOPES**
BRULE, WESTBROOK, SOUTHAMPTON, KENNEY
BRULE, PORTAPIQUE, FOLLY
BRULE, SPRINGHILL, POBSON

**UNDULATING TO ROLLING STONY SANDY LOAMS - MAJOR PROPORTION IMPERFECTLY TO POORLY DRAINED**
LYDGATE, FORT HEBERT, ROCK KNOBS, PEAT
DANESVILLE, BAYSWATER, LIVERPOOL, RIVERPORT
HUMMOCKY TO GENTLY ROLLING STONY SANDY LOAMS AND LOAMS - WET AND DRY

**SOILS IN A COMPLEX PATTERN**
YARMOUTH, DEERFIELD, LIVERPOOL, MERSAI, DANGERFIELD, BRIDGEWATER
THOM, GIBRALTAR, ROCK KNOBS, PEAT

**STRONGLY UNDULATING TO MODERATELY HILLY STONY SANDY LOAMS AND SHALY LOAMS**
GIBRALTAR, HALIFAX
HALIFAX, BIRCHHILL, GIBRALTAR, RAWDON
HALIFAX, MERSEY, GIBRALTAR, PEAT
KIRKHILL

**ROLLING TO HILLY STONY SANDY LOAMS AND GRAVELLY SANDY LOAMS - LARGE PROPORTION OF STEEP STONY SLOPES**
OREGON, NYSSEN
BERWICK, SPENCER, PAISRENCE
THORP, SHERBROOKE, BARNEY
ROSSBURN, ROCKVILLE, THORNVILLE
BASS RIVER, MORTENHOUSE, LAKE, ROCK OUTCROP

**UNDULATING TO ROLLING LOAMS AND SANDY CLAY LOAMS - LARGE PROPORTION OF DRUMLINS**
BRIDGEWATER, WOLFVILLE
WOLFVILLE, MORRISTOWN

**GENTLY UNDULATING TO ROLLING CLAY LOAMS**
FALMOUTH, WOLFVILLE
MIDDLETON, LAWRENCETOWN

**GENTLY UNDULATING TO ROLLING CLAY LOAMS - LARGELY IMPERFECTLY DRAINED**
QUEENS, CANNING
DALHOUSIE, SHREWSBURY
QUEENS, WOLFVILLE, HANTSPORT

**ROLLING TO HILLY GRAVELLY CLAY LOAMS**
WOODBOURNE, MILLBROOK

**UNDULATING TO DEPRESSIONAL SANDY LOAMS - MOSTLY POORLY DRAINED**
MASSTOWN, ECONOMY, PEAT
ASPOTOGAN, PITMAN, ROSEWAY, PEAT

**GENTLY UNDULATING TO DEPRESSIONAL CLAY LOAMS - POORLY DRAINED**
KINGSVILLE, DUBRIDGE
JUNKINS

**ORGANIC SOILS**
PEAT, MUCK

virtually complete. This project has been carried on by the Photogrammetry Division of the Nova Scotia Research Foundation with the object of producing a map showing glacial surface features such as drumlins, kames, eskers, outwash plains, till areas, raised beaches, old shorelines and old lakes. This mapping project has already proved of great value to the construction industry; it should prove useful to all who wish to obtain aggregate for construction, or to know upon what they are building. Studies to date indicate that the effects of glaciation on the bedrock were relatively small. The loose or partly decayed top of the rock masses was removed as by a bulldozer, but the resultant debris was not transported for any great distance. It can be definitely stated that transport in many cases was about a mile, or at the most two miles.

This is particularly clear in western Nova Scotia, in Lunenburg County, where the slates of the Meguma Series yielded clay debris which was removed, transported about a mile, and then overridden by the ice to form numerous drumlins. In areas of softer rocks, such as the lowlands of Cumberland County and in Pictou County, the mantle of displaced debris is more evenly distributed but shows about the same displacement. Erratics are widely distributed, but with few exceptions, make up less than ten per cent of the glacial mantle. A notable exception is the basalt of North Mountain, which is widely distributed in large quantities over southwestern Nova Scotia. This is readily explained on the basis of its exposed position on the top of the cuesta ridge of North Mountain and its highly jointed character. Another exception is the lobe of reddish material found in central Nova Scotia in Halifax, Lunenburg and Hants counties. This is believed to be the result of the erosion by the ice of one of the continental glaciers of a mass of soft Triassic shales, which once occupied what is now Minas Basin. These shales were unprotected by a basalt lava cap and were thus exposed to erosion by the advancing ice sheet which came down from the north west.

## CONCLUSION

It is concluded that the effects of continental and local glaciation were relatively slight and that transportation over long distances of large amounts of material derived from the weathered bedrock did not take place. It can be said in general that the soils have been derived from the bedrock on which they rest by a process which started with the disturbance of the weathered surfaces to form a mantle of glacial drift, with little transport, followed by weathering of the glacial material to form soils.

## ACKNOWLEDGEMENTS

The writer wishes to acknowledge the help given by the Soils Group of the Joint Dominion Provincial Soils Survey of the Canada Department of

Agriculture and the Nova Scotia Department of Agriculture and in particular, the help of Dr. Bruce Cann, who most generously offered the use of the generalized soils map of the province. The writer also wishes to acknowledge the opportunity provided by the Nova Scotia Research Foundation to carry on the research in Pleistocene geology, bedrock geology and physiography which has made this study possible.

## BIBLIOGRAPHY

1. CAMERON, H. L. (1948). Margaree and Cheticamp map-areas, Nova Scotia. Geol. Surv., Can., Paper 48–11.
2. ———— (1949). Faulting in Nova Scotia. Trans. Roy. Soc. Can., Series III, *43*, Sec. IV: 13–21.
3. ———— (1949). Sub-sea faulting in the Lismore area, Nova Scotia. Trans. Can. Inst. Min. Metall., *52*: 87–8.
4. ———— (1956). Tectonics of the Maritime Area. Trans. Roy. Soc. Can., Series III, *50*. Sec. IV: 45–51.
5. CANN, D. B., and HILCHEY, J. D. Reports of the Nova Scotia Soil Surveys: no. 5, Hants County (1954); no. 6, Antigonish County (1954); no. 7, Lunenburg County (1958); no. 8, Queens County (1959).
6. CANN, D. B., and WICKLUND, R. E. (1950). Report No. 4. Nova Scotia Soil Survey (Pictou County).
7. GOLDTHWAIT, J. W. (1924). Physiography of Nova Scotia. Geol. Surv., Can., Mem. 140.
8. WHITESIDE, G. B., WICKLUND, R. E. and SMITH, G. R. (1945). Report No. 2. Nova Scotia Soil Survey (Cumberland County).
9. WICKLUND, R. E., and SMITH, G. R. (1948). Report No. 3. Nova Scotia Soil Survey (Colchester County).

# ORGANIC TERRAIN

## Norman W. Radforth, F.R.S.C.

WHERE THERE ARE LANDS covered with vegetation growing in peat the condition is designated by the expression Organic Terrain. An understanding of the phenomena associated with this terrain requires application of knowledge of at least four topics: micro- and macro-palaeobotany, oecology and physiography. In the present paper emphasis will be placed upon the significance of physiographic relationships in organic terrain studies. To accomplish this it will be necessary to elaborate first on the other subjects mentioned, for the natural synthesis of the material of which the terrain is composed relates fundamentally to them; and physiography, if it is to have any influence, will directly or indirectly control terrain structure.

The work of Von Post in which microfossils are utilized in assessing past climate and forest history is classical. Among the contemporary contributions, that of Faegri and Iversen (2) is perhaps most widely known by palaeobotanists. To perform micro-analysis, fossil pollen and spores are retrieved from core samples of peat. Usually major emphasis is placed on fossil tree pollen; the objective is not usually concerned with the history of the peat, but with sequence in the microfossil record preserved in the peat. The evolution so revealed is of forests that have contributed the broad environment of the organic terrain.

Thus it will be appreciated that, fruitful as this work is, it sheds little light on peat organization; indeed the author has seriously questioned whether peat does reveal organization. To claim that organic terrain is organized requires more extensive examination because, even if peat were organized, it would be only one of the factors included in the expression organic terrain.

In flights over organic terrain observers are impressed with the apparent lack of cosmic arrangement (Figure 13). The admixture of pattern seems unaccountable. There is little certainty that any given pattern is an expression of natural events and not a mere configuration of subjective inspection and therefore artificial.

The observer at ground level has even less assurance, for he cannot extend beyond the limits of micro-environment. Where there is change in the state of the terrain, he cannot assess its value in relation to the surrounding area. He notes differences in micro-topography, vegetal cover and peat structure, but cannot be certain that any natural relation is expressed through them. This is the most important difficulty for an interpreter of organic terrain, because if there is no apparent organization he cannot make any classifications.

There was perhaps a suggestion that cosmic relations might be expressed in organic terrain inasmuch as contemporary vegetation has an ordered arrangement. It could seem reasonable that the fossil legacy of plant cover might also express laws of order. At the McMaster University Muskeg Laboratory a start was therefore made to segregate and identify macro-fossils in the peat and to study their sequence in relation to their vertical and lateral occurrence in selected peat deposits. Procurement of reliable evidence was difficult. Often macrofossils were fragmentary and could not be identified. Frequently they overlapped; a given fossil would be preserved on three planes and surround others that were not; or some had fallen into place and others had not, depending often on the size of the fossils. It was found also that interpretations to predict an ordered state failed. For these and other reasons the macrofossil study was temporarily abandoned.

Attention then had to be directed to the microfossils. These were discrete and their features were diagnostic. They were all of much the same size and whatever differences there were did not appear to affect their random distribution in sedimentary deposits. The fossils were so small that they would be "trapped" in significant numbers very close to where they came to rest on solid matrix.

That some of them might have been carried by wind or water from beyond terrain under study was accepted. Those that could have come from afar could usually be differentiated from those that originated from plants that typically contributed to the structure of the peat. Unusually frequent microfossils that could have come equally well from a distant plant or from one growing in the peat were qualitatively weighted as probably originating from the plant in the peat.

The results of investigations on microfossils are reported in several accounts (5, 7, 11). No attempt will be made to review them here, but it will be necessary to refer to some of the conclusions of former work in order to describe organic terrain as a whole. When *in situ* plants are emphasized in microfossil sequence it assists comparison of one example of terrain with another, if geographical delineation is first established for the areas in question. Microfossil analyses of samples taken from such adjacent areas show that four aspects of comparison arise: (1) Microfossils may show similar frequency relations for successive depths of peat for samples from different sites in given areas. (2) Microfossils may show similar frequency relationships for successive depths of peat at given sites but the frequency history may differ among sites in a given area. (3) Microfossils may show dissimilar frequency relationships for successive depths of peat for a selected site in a given area; if so, the dissimilarity usually suggests successional trend in the history of plants at the site. (4) Where dissimilarity in relation-ship of frequency exists at a given site, the pattern may or may not be the same for other sites where dissimilarity is also expressed.

If consideration is now given to the history of the fossil vegetation compos-ing the peat as conveyed through microfossil sequence, it seems reasonable to claim that since the last retreat of ice, vegetal composition *en masse* may have experienced no primary change over a significantly large expanse of land. This situation may also obtain even though composition in one major area may differ from that in another.

There is also reason to claim that where succession is exhibited in vegetal composition, this condition may be interrupted over an expanse by contrasting kinds of vegetal composition each of which primarily shows constancy within itself. It will be appreciated that, if such interruptions occur, it may be the same kind of vegetal composition that interrupts. Finally, it may be reasoned that a successional trend in vegetal composition whether it is interrupted or not may give way to other kinds of trend.

It is tempting to declare these conclusions as acceptable principles on the basis of evidence already at hand (5, 7). The writer prefers, on the other hand, to regard them as proposals within a working hypothesis on the hope that as cases arise secondary refinements will accrue, for example, vegetal history as expressed in the Copetown Bog (11). As the evidence increases, understanding becomes also more complete on relations between fossil vegetal composition and contemporary vegetal cover; additions to the basic hypothesis may be in the future useful extensions to the basic principles.

In the meantime there is no need to postpone exploitation of the situation that organization in fossil vegetal composition exists as revealed through microfossil study. There is encouragement for the interpreters of organic terrain arising out of the fact that the microfossil relationships indirectly infer *in situ* vegetal relationships because, if constancy or inconstancy are assessible, distributional analyses of these conditions will account for the variation in macrofossil arrangement to which reference has already been made (3).

Although it is not easy to establish that predominating microfossils signify the same species of macrofossils at equivalent depths at given sites, it is easy to discern that characteristic associations and frequencies of microfossils denote the presence of corresponding kinds of plant structure in the peat. This encouraged investigation of variation in plant structure in peat deposits. It has already been suggested that, on an arbitrary basis, there are sixteen major categories of peat that are visually distinguishable (7). All differ enough structurally to suggest that each would probably be distinguishable from the others on the basis of mechanical properties. Unfortunately, no system of instrumentation has yet been devised that would be useful in substantiating this claim.

Qualitative differences in peat structure are therefore so far recognized on a visual basis. Structural variation is in part a function of fibrous and non-fibrous constitution. In peats that are mainly fibrous, size, arrangement, density and degree of woodiness of the fibres play a part. In peats regarded as non-fibrous, aggregations of granules predominate; they lack specified shape (amorphous) and are associated to produce visibly different densities. Usually the granular is mixed with the fibrous condition.

It will be appreciated that with so many variables, the structural constitution of peat could be so varied as to form many more categories than the sixteen selected by the writer as standards. It cannot be denied that secondary categories exist. In nature the situation is such that the commonest associations are in fact the sixteen chosen ones and of these several are very common. Perhaps the reason for so few categories is the same phenomenon that controls vegetal composition as suggested by microfossil composition.

An example might clarify the relationships expressed. When sphagnaceous microfossils occur in relatively large proportions they are accompanied by relatively large frequencies of ericaceous microfossils. The proportions along with those of the other microfossils are approximately the same at intervals throughout the depth of the peat. In these circumstances the primary category of peat that is representative of the structure is number 14* (non-woody and woody fine-fibrous held in coarse-fibrous). There are variations. If sphagnaceous microfossils are very high, and ericaceous ones very low, the peat structure category may be 11 or 4 (11, pp. 39–40). The latter is relatively high in amorphous granular matter and both contain more non-woody fine-fibrous matrix than do samples of category 14.

It must be emphasized that a given microfossil conspectus has equivalence in peat structure and not usually in peat macrofossil species or even genera. Thus, constitution for a given sphagnaceous-ericaceous complex although nearly constant structurally may be low in *Sphagnum* and high in *Hypnum* macrofossils. The woodiness may be owing to the presence of *Ledum* in the peat but it is often the case that *Kalmia, Vaccinium, Andromeda, Empetrum* etc., predominate instead of *Ledum* and in various proportions.

Attention may now be directed to the contemporary vegetal cover with the thought in mind that it is the next layer which will eventually contribute to the organized micro- and macro-fossil constitution. Marked relationship is expressed between the microfossil constitution and vegetal structure on the surface (4, p. 11), at given sites. Thus the high sphagnaceous-ericaceous proportion in a microfossil conspectus corresponds at the surface to certain structural classes of vegetal cover. It does not correspond necessarily or even usually to constant genus or species associations. Just as in subsurface constitution, vegetal relationships with the microfossil conspectus are structural not generic, so in surface constitution the relationships with the underlying microfossil complex are also structural and apparently not generic.

It has been explained elsewhere (9) that assessment of surface cover if attempted on a species basis may lead to bewilderment because too many species transgress micro-environments. Consequently, differing species or admixtures of species share similar environments. On the other hand, if cover is assessed on a structural basis (stature, form, texture, woodiness *v.* non-woodiness etc.) it characterizes environment. On the structural bases nine classes of cover are necessary to categorize the total effect of

*Classification numbers as given in reference (5).

structure. They have been symbolized by the first nine letters of the alphabet (A to I inclusive) (4). Classes C and G are rarely significant.

Cover analysis requires the application of several fundamental steps:

1. The areas of organic terrain under investigation are delineated.

2. Cover formulae are determined for each area. This is accomplished by noting the associated classes (4, p. 8), and arranging their symbols in order of predominance. The symbol offering maximum cover over-topping the others is placed at the left of the formula. The symbols for other classes are added in descending order of significance. If a class shows cover value estimated at 25 per cent or less it is arbitrarily excluded from the formula. Thus, a formula may have one, two or three symbols in it but not four. Usually it has two or three.

3. Usually formulae show discrete geographical boundaries and these often coincide with the delinations set by the interpreter. If this does not obtain, adjustments in initial delineation should be considered because coincidence often proves convenient when data are applied for mapping and other purposes.

The reader may be apprehensive about the possibility of so many formulae arising even in the knowledge that classes C and G are infrequent. Experience has demonstrated that actually only a few of the possible number occur in nature. Of these, several are very common and if they do not define very large areas they recur very frequently. It sometimes arises also that, if there is change in cover formula, it is because the relative predominance of two classes requires the interchange of two class symbols in the formula.

Application of the interpretive method shows that formulae are associated in families. With practice, the interpreter automatically assesses in terms of families in his initial superficial survey especially if the area under investigation is several square miles in extent. The commonly occurring formulae are arranged by families in Table I. Note that for some families only one formula is considered as common. Examination of Table I will show that family groups are discernible owing to the prominence of a single class in the cover. It will be appreciated on the other hand that secondary relationship exists among families. This is demonstrated in Table I where it is seen that EI is a component of FEI, BEI and AEI. Primary and secondary

TABLE I

FAMILIES OF COMMON COVER FORMULAE

| Class | Formulae |
|-------|----------|
| A | AH   AEH   AEI   AFI |
| B | BEI  BFI   BDF |
| C | Not Contributory |
| D | DFI |
| E | EI |
| F | FEI  FI |
| G | Not Contributory |
| H | HE |
| I | Low Predominance |

FIGURE 1.    Cover formula AH

FIGURE 2.    Cover formula AEH

relationship will be appreciated through comparison of Figures 1 to 12 which show examples of the cover formulae listed in Table I.

To appreciate analytical procedure fully it is suggested that the reader examine the photographs with the aid of the key data published elsewhere (4, p. 6, Table I). Intra- and inter-family relationship among cover formu-

FIGURE 3. Cover formula AEI

FIGURE 4. Cover formula AFI

lae can also be appreciated to a limited extent by a comparison of Figures 1–12.

The more typical condition of AEI is shown at the upper right of Figure 3. In this area, class A is relatively denser than it is in the foreground, E has a more prominent stature and the cover generally is more like the typical condition of BEI (Figure 5) except that the trees are consistently much larger. In the foreground of Figure 3 and occupying most of the photograph the total canopy of class A and that of E compete for prominence. If A and E were interchanged in the formula there would be little objection. Class I though present generally in Figure 3 is by reason of its stature indiscernible. In the background, in front of the more typical AEI, a patch of class H appears and is noticeable by reason of its light colour. This will suggest that in the lowest layer of the living vegetal cover, there is competition for prominence between classes I and H. Then too, there are cases where class E is negligible, for example, Figure 1 AH. Thus *intra*-family relation is expressed when there is change in class predominance within a type formula, AH AEI AEH. *Inter*-family relation- ship is expressed when the foremost class denoting family is occupied by another family symbol e.g. BEI, Figure 5, in contrast to AEI, Figure 3, typical condition upper right. It can also be expressed through the fact that EI is common to both formulae.

Consideration of cover formula BEI (Figure 5) will emphasize similar situations based on family relationships. There are other principles con- veniently expressed by this example. Note that the combination BH does not appear as a common member of the family. Indeed it is questionable that it exists, according to the writer's experience, although AH is common enough within its family. It should also be emphasized that in BEI class B is not a juvenile condition (as to stature) of class A; its characteristic stature is sustained. Finally, there is clear indication in the photograph Figure 5 that class B is composed of more than one species of tree. Thus, it is reasonable to suggest that *intra*-family relationship may be expressed differently in different families despite inter-relationship, that features (e.g., stature) defining class may be sustained and are not directly attributable to species (genetical) but to environmental conditions.

In the comparison so far it will have been appreciated that structural composition has not dealt with major items of contrast. Because of certain aspects of family relationship possibly AH (Figure 1) has been the only cover formula suggesting marked aberrancy (cf. Figures 1, 2, 3, 5). Con- sider now Figure 4, cover class AFI. Only the uppermost layer of cover is woody. The effect is such that the observer senses at once the marked con- trast with the condition seen in AEH or AEI, Figures 2 and 3. This indicates that within a family contrast in the appearance of the cover can be extreme.

Inter-family relationship is shown in a comparison between AFI and BFI, Figures 4 and 6 respectively. Sometimes intra-family relationship shows cover contrast more extreme than that noticed in inter-family relationship. Some explanation is required for cover formula BDF, Figure 7. On the bank of the river, DFI forms a ribbon separating BDF in the background from the water. This is an example of how demarcation between cover formulae and, in this case, cover family can be abrupt.

A typical condition of DFI is shown in Figure 8. Comparison between it and the example appearing in the foreground of Figure 7 suggests dissimi-

FIGURE 5. Cover formula **BEI**

FIGURE 6. Cover formula BFI

larity on casual examination. In the one case, Figure 8, foliage is present on the plants of class D whereas for the same class in Figure 7 it is absent. Hence the difference is not fundamental with respect to the terms of classification. Seasonal variations in cover appearance must be accommodated and reconciled by the observer during analysis, if they are thought to be significant in facilitating further interpretation.

FIGURE 7.   Cover formula **BDF**

FIGURE 8.   Cover formula DFI

The absence of tall cover layers provides a further basis of contrast. This is noticed in a comparison between cover formulae AEH, Figure 2, and EI, Figure 9. The latter is the only major example typifying the E cover family. The photograph Figure 9 shows the vegetation at close range to indicate that cover class F is present. In this case F is estimated at about

FIGURE 9. Cover formula EI

FIGURE 10. Cover formula FEI

twenty per cent of the cover and for this reason does not warrant inclusion in the formula. Figure 9 shows the presence of class I on the face of the mound seen about half an inch from the base of the photograph right of centre. Where the lower edge of the photograph cuts through the mound immediately in front of the one displaying class I there is a small patch

FIGURE 11.   Cover formula FI

FIGURE 12.   Cover formula HE

of cover class H. It obviously lacks significance so far as cover designation is concerned.

When class F is predominant and associates with classes E and I the condition usually appears as shown in Figure 10, a photograph of cover formula FEI. In this example classes E and I are nearly equal in prominence and much of both is hidden from view by class F. The dark-coloured isolated portions of the class E component are easy to see. Some class I cover can be detected at the lower edge of the photograph.

The relative importance of class H is apparent in cover formula HE, Figure 12, the light-coloured broken background coexisting with the dark which is class E. Towards the back of the picture, behind the dwarfed tree, is an expanse of formula FI spreading left and right and forming an abrupt margin for formula HE. When formula FI is inspected at closer range it appears as shown in Figure 11. Its relationship with formula FEI, Figure 10, can be appreciated and yet there is good evidence of contrast between the structural features displayed by the two formulae.

The comparison of all the photographs may have demonstrated enough to suggest that pure classes of cover seldom exist by themselves. The significant references for cover classification are the cover formulae. As this system of vegetal cover classification has come into use the question has arisen as to whether the formulae rather than the classes could be regarded as the basic items of reference. This would be acceptable were it not for the difficulty of describing the cover for each formula. Thus whether it be in performing an analysis of the composite cover or in explaining its synthesis, reversion to use of classes as elementary components seems essential. This is not to suggest that formulae have no natural significance. In other words, although they have convenient practical value for interpretive procedure, they are not artificial or without biological significance.

In terms of area and distribution of cover, random size and arrangement seem to obtain. There is good reason to doubt that this is the real situation as will be explained later when other conditions of the terrain are examined. In the meantime, it is difficult in either circumstance for an untrained observer to classify cover and to recognize possible trends over very large areas because of the apparent random condition.

Attempts to comprehend cover arrangement have been made largely by aerial inspection. As one proceeds to higher altitudes over a given "target" of organic terrain, structural patterns in the cover suggest themselves just as they do on the ground but with a different basis for formulation. The elements of structure that define cover class on the ground are not all discernible even at low altitudes. From the air the eye is attracted instead to zone delineation of structural entities that embrace and characterize masses of vegetation; generally, the higher the altitude the larger the masses.

At altitudes of 30,000 feet, the patterns which categorize cover are therefore quite different in appearance from those on the ground and are still apparently arranged at random. Areas of a given kind of pattern also differ considerably in size as do component cover formulae seen from the ground. These conditions will be appreciated in an examination of Figure 13, an aerial photograph taken at approximately 30,000 feet over terrain that is nearly all organic. The photo shows mainly marbloid, reticuloid and terrazzoid patterns but small areas of stipploid and dermatoid patterns are also present.

The analysis of vegetal cover as seen from the air has been discussed in other accounts (8, 9, 10). The distinctive areas noted in Figure 13 are now known as "air-form patterns." They are significant for several reasons, but the observer is reassured of their validity in the realization of a single reason, that these patterns recur across Canada. In some parts of the country, certain patterns are characteristically higher in size and frequency than others. The writer has not had the opportunity to examine air photos from all the provinces but he feels confident in predicting that, with the exception of Prince Edward Island, all air-form patterns will appear in one proportion or another in all the provinces.

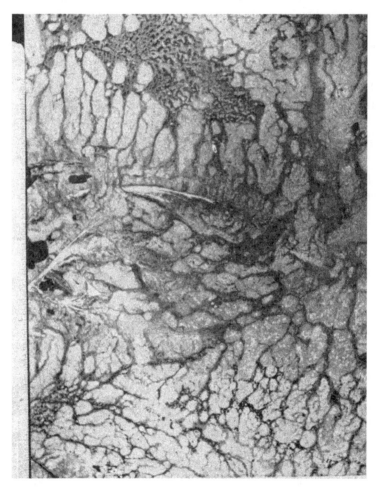

FIGURE 13. Aerial photograph from 30,000 ft. of almost completely organic terrain. Scale: 1 inch to 1 mile.

The recognition and naming of air-form patterns was not achieved by either reconnaissance at 30,000 feet or examination of 30,000 feet photos. Aerial reconnaissance investigation began at altitudes of 150 to 300 feet, with the aid of Sonne strip photography over areas of organic terrain already classified by vegetal cover formulae determined at ground level. At this low altitude all diagnostic features defining cover at ground level could be fairly easily seen. But at this same altitude air-form patterns began to emerge and persist in air photos recorded at all altitudes up to about 1,000 feet. Thus ground level character in the cover, integrated into 1,000 feet air-form patterns, although often not discernible, could be accounted for with the aid of the low altitude photos or flights and therefore subsequently interpreted without these aids. This method of extrapolating ground pattern (cover formulae) was used eventually to altitudes of 30,000 feet.

When this altitude was reached analysis showed that the air-form patterns now characterizing 30,000 feet experiences are the logical ones. For example, a given air-form pattern, when analyzed, facilitates inter- and intra-family relationship among the cover formulae represented. In fact it tends to confirm the existence of families of cover formulae and thus to justify their use not only as valid artificial diagnostic entities but also as natural ones.

More complete acknowledgement of a claim that natural values exist as represented by cover formulation arises when formulae are compared with microfossil constitution of the peat beneath them (5, 11). With dimensions of time and depth available in the microfossil examinations and applicable in accounting for the origin of contemporary cover, the existence of organization becomes more evident.

As developments in organic terrain interpretation now stand, the wide assortment of cover expressed in Figures 1 to 12 is accompanied by equivalent assortment in microfossil conspecti. Because the latter also show relationship with macrofossil constitution, it is reasonable to judge macrofossil composition (peat structure) directly from vegetal cover analyses. If this idea is extended to accommodate aerial interpretation subsurface constitution in the peat can be estimated by inspection of the aerial photographs taken at 30,000 feet (9).

It is seldom the case that organic terrain occurs without living cover unless the peat is laid bare by fire or wind action. Therefore for all practical purposes classification using living cover can always be achieved. As may be expected, there is a minority of examples that are more difficult to classify than most. This arises when two (seldom more) cover formulae occur as an admixture, or when arbitrary delineation of an area inconveniently transects cover components to give secondary importance to one formula whereas in fact it predominates in the environment of the area. Or, finally, analysis may be required on the northern rim of the subarctic where there is scarcity of tree cover and cover differentiation is therefore difficult at a distance.

Under these circumstances, it is usually helpful to the interpreter if he resorts to classification by topographic features. There are characteristic micro- and macro-topographic elements of reference, many of which are related closely to cover formulae and therefore indirectly to subsurface and aerial classification and interpretation. These features have already been named, symbolized and classified (8, 10) and the writer does not wish to elaborate directly on their relationships here. They have indirect significance, however, in the present discussions because they are useful in an understanding of the physiography of organic terrain to which consideration may now be given.

In spite of the fact that complete classification and interpretation of organic terrain is not simply a function of cover and subsurface peat constitution, and that physiographic inference is made in the definition of muskeg (8, p. 4), there is little in the literature that offers any explanation that establishes or classifies possible physiographic connections. In addition to the writer's contribution on topographic difference, and the importance of size and shape of air- and ground-form pattern (8, 10), there is the classical reference to high and low moor peat land conditions. Recently the work of Miss Allington (1), sponsored by the Defence Research Board of Canada, has made a useful contribution to an understanding of some terrain configurations and there is indication that Russian investigators have contributed in this connection (3).

The writer is interested here in demonstrating the basis on which physiographic phenomena present themselves generally in organic terrain interpre-

---

## TABLE II

### Mineral Terrain Conformation—
### the Foundation of Organic Terrain

| | | Axial Foundation | Symbol | Areal Foundation | Symbol |
|---|---|---|---|---|---|
| DIAGNOSTIC | Single Features | Straight, Abrupt Curved | ▬ ∿ | Area Featureless | Ω |
| | | Sinuous, Angular | ∿ | Area Pitted | ⊽ |
| | | Sinuous, Smooth | ∿ | Area with Clefts | ⊽ |
| | | Reflexed | ᶯ | Area with Shadow Depressions | ⌣ |
| | | Discontinuous | ‒‒ | Area with Folds | ⌐ |
| | | | | Area Contorted | @ |
| | Compound Features | Barred Complex | ≕≕ | | |
| | | Concentric Curves | ≋ | | |
| | | Paired Sinuous | ≈ | | |

tation and classification as these have evolved in his own experience. Field observations made on the subject over the years are sufficient in number and suggestive enough to enable hypotheses to be formulated. One simple observation is that mineral terrain conformation is often reflected through the organic mantle; sometimes it coincides with distribution patterns of cover formulae.

The kinds of geomorphological features that are involved in this relationship are shown in Table II. It will be noted that reference to the various features is by description rather than by name. At this early stage in the study, it would be unfortunate to promote misunderstanding by innocent misuse of geomorphological nomenclature. The observations relate to shape, size and distribution of the physiographic entities but not to their academic definition or genesis. Thus, the significant mineral terrain features are in the first analysis either axial or areal. If the former, they may be single or compound (Table II, left hand column). There is justifiable contention that the compound features described are really areal, and therefore inaptly classified. On the other hand, those features listed as "areal" show no certain relationship to the "axial," whereas the compound and single axial features are interrelated as the corresponding symbols suggest. All of the features appear by reason of a difference in elevation to that of the surrounding terrain. In a few cases (e.g. pitted, clefts) it is the terrain that surrounds the feature that is elevated in the relative sense.

There are common cover formulae which assist in characterizing these mineral terrain features. These are cover formulae AH, AEH, AEI, ADE and HE for the axial type of foundation and AEI, EI and FI for the areal foundation. There is a little overlap in that AEI is common to both series of geomorphic features. Overlap would be emphasized if the less common cover formulae were added. Thus the formulae are not aids to conclusive identification of features beneath them; they are only partial aids. In any case, the geomorphic feature can be often identified by its inherent conformation. The formulae really show that natural relationship exists between geomorphic features and vegetal cover (and therefore subsurface constitution). An example illustrating this principle is shown in Figure 1. The curved light-toned area is made up of sites of AH and HE. It defines the geomorphic configuration. Beyond, the cover is identified by the formula AEI. The configuration is axial curved and is caused by relatively higher elevation in the mineral foundation of the organic terrain. If there is relatively high elevation, the cover is almost invariably AH; if low, it is AEH. Where there is frequently no tree growth, as happens in the north, the cover is expressed by HE, and the configuration is surrounded by FI.

Although recognition of mineral terrain configurations beneath organic terrain is a comparatively new study, the interpretation of the configurations lacking an organic mantle is probably as old as geology. More recent than either of these is the practice of identifying, classifying and interpreting configurations that arise within the organic overburden producing conformation that therefore would not arise if organic terrain were not present. This kind of conformation is described in Table III.

TABLE III

ORGANIC TERRAIN CONFORMATION

| Axial diagnostic form | Areal diagnostic form |
|---|---|
| Straight, sides Parallel | Featureless |
| Tortuous, Nodular | Elevations, smooth border |
| Crescentic or Circular | Elevations, jagged border |
| | Perforated |
| | Reticulated |

As with mineral terrain the configurations are either of an axial or an areal category. The axial features are markedly narrower than those that characterize the mineral sub-layer. The common order of width of the axial configurations is about five feet. The fact that they are differently shaped suggests the influence of a physiographic differential influencing the local rate of accumulation of the peat.

One of the conformations listed under "areal" Table III shows no elevated configurations and is therefore designated as featureless. Sometimes an expanse of this kind becomes interrupted by small depressions (perforations in the surface). It should be emphasized that these do not extend into the mineral sub-layer and are therefore presumably a function of the dynamics of peat formation. Other configurations listed under "areal" Table III, are discernible from 30,000 feet (Figure 13). In the photograph smooth- and jagged-bordered elevations associate to form the typical marbloid air-form pattern and the reticulated condition shows up as a meshwork.

The common cover formulae occurring on the configurations peculiar to organic terrain are BDF, DFI and FI on the axial ones and EI, HE and FI for the areal ones. Given formulae are not specific for given configurations but wherever a configuration occurs the vegetation covering it differs in structure from the surrounding kind. For interpretive work it is useful to know that cover formulae relating to geomorphic features of the mineral sub-layer differ on the whole from those prescribing for features of organic terrain origin.

Conformation generally is further complicated by a seasonal factor which is most marked in permafrost country and in subarctic lands immediately to its south. Configurations are noticed as incorporated features of the organic terrain and are contributed entirely by the presence of ice. Reference to subsurface ice constitution in organic terrain will be found in some of the writer's earlier work (6, 8). Work in the field has revealed ice conditions suggested by the descriptive terms listed in Table IV.

In some cases the incorporated ice mass is continuous and recedes into the surface of the terrain approximately equally at all places. This condition

TABLE IV

|  |
|---|
| Featureless |
| Fissured |
| Perforated |
| Isolated Masses |
| Polygonous |
| Knolled |
| Sculptured |
| Ridged |

is said to be featureless. In late spring, in some examples of terrain, fissures are found and in others holes occur in the imbedded ice. The ice in very late spring sometimes shows the effect of unequal melting, and large isolated masses or cakes of ice with tapered edges remain behind for periods extending into the summer months.

When the ice is continuous, it is not always featureless. Sometimes it conforms to a polygonal pattern that involves elevated areas of ice five to fifty feet in diameter separated by depressed boundaries (cf. figure iv). Often its surface is covered with dome-shaped extrusions that are usually contiguous at their bases. This condition is now known as ice-knolling. In other cases the surface of the ice is sculptured with confluent depressions of various shape and depth. Finally there are examples where the ice surface is ridged rather than knolled. If ice configurations are comprehended on a distributional basis it will be appreciated that among the physiographic factors ice alone is responsible for much micro-topographic variation. That the condition is never static emphasizes the fact that the mechanical properties of organic terrain are specialized as well as inherently complex.

Turning to a consideration of vegetal cover as viewed in relation to subsurface ice, it may be stated at the outset that all the common cover formulae, and others less frequent, are to be found pertinent to one or other ice condition. For the featureless state, the commonest cover formula is FI, but this is also characteristic for the perforated examples. It is fairly consistently the case that, in covers where class I in quite local situations displaces F, a perforation develops in the ice beneath the surface. The size and shape of the perforation corresponds roughly to that of the patch of I. Perforated ice is also found beneath DFI cover; the holes occur among units of class D.

It is not usual to find isolated masses of ice beneath FI or DFI cover. The ice thins out generally to virtually nothing as the perforations enlarge. Isolated ice masses are found beneath cover class HE especially when H and E are nearly equivalent in prestige. Less commonly they are found beneath EI, but the characteristic subsurface ice condition for this kind of cover is the knolled one. The polygonous condition is found beneath cover formulae HE (commonly but not generally), FEI and occasionally in EI. To the writer's knowledge it arises under neither DFI cover nor the A and B families cover formulae. Beneath the latter cover formulae, the knolled condition occurs but it is so common under formulae AEI and BEI that it may be said to characterize them. It is universal for EI cover and persists to mid-summer as far south as Kapuskasing, Ontario.

The sculptured effect on the subsurface ice reflects two orders of significance. One is observed where there is a change at the surface from one kind of vegetal cover to another. The sculpturing shows greater amplitude of elevation when peat plateaux (8) are involved anywhere in the total expanse of terrain. The other order of sculpturing occurs beneath a predominating cover formula, especially EI, when the formula is interrupted at frequent intervals by smaller secondary patches of another cover especially FI with EI. Subsurface ice ridges are usually incorporated into axial configurations originating in the organic terrain. They are more frequent beneath formulae EH and HE. When they occur beneath EI they are usually short or discontinuous. Ridged ice never appears beneath the formulae of A, B, D or F families to the writer's knowledge.

There are also obvious effects of physiographic influence caused by water (Table V). Diagnostic form for water-imposed conformation is axial and areal as it is for the configurations listed in Tables II and III. These features

TABLE V

WATER IMPOSED CONFORMATION IN FOUNDATION
OF ORGANIC TERRAIN

| Axial diagnostic form | Areal diagnostic form |
|---|---|
| Tortuous—Angular | Featureless |
| Tortuous—Smooth | Angular Expanse |
| | Rounded |
| | Dendritic Course |
| | Anastomosed Course |
| | Paralleled Course |

are easily recognized from the air at any altitudes up to 30,000 feet and the descriptions (Table V) are self-explanatory. The depression relating to the concentration or drainage direction of the water may be primary, that is, arising originally in the mineral terrain, or secondary, formed as a result of barriers generated in the organic terrain itself. Whether axial or areal, the drainage course or reservoir, as the case may be, is covered with vegetation.

Cover formulae pertinent to the water-prepared features are characteristically FI and DFI. When the former obtains, it sometimes happens that the formula transgresses the boundaries of the configuration. The water zone is invariably darker in colour once spring growth is established. Though the cover formula does not change, demarcation of the physiographic phenomenon is therefore clear.

Among the axial examples, there is greater likelihood of variability in cover when the drainage course is angular than when it is smooth. With the angular condition the drainage gradient is characteristically steeper and

| TABLE VI | | | | | | |
|---|---|---|---|---|---|---|
| Cover families and Order of Importance of Associated Terrain Conformations | | | | | | |
| Families | A | B | D | E | F | H |
| Order of Importance of Conformations | Mineral Ice Water Organic | Mineral Organic Ice Water | Water Organic Mineral Ice | Organic Ice Mineral Water | Organic Mineral Water Ice | Ice Mineral Organic Water |

the cover formula will more probably be DFI than FI. When the course is tortuous-smooth (Table V, col. 1), although the cover is usually symbolized by FI, the formula DFI may be present as a fringe on both sides of the FI cover.

Among the areal categories, the featureless condition is invariably large and not limited abruptly by sudden change in elevation or cover formula. Usually the vegetal cover is FI. When a reservoir occurs covered with vegetation and has marked angular boundaries, it is spoken of as an "angular expanse." Unless the cover formula is DFI, the abrupt demarcation at the boundary is because of sudden relative increase in elevation. If the increase is quite sudden, the cover formula changes almost at once from DFI to AEI; if it is gradual it changes in the outward direction from DFI to BFI, BDE and finally AEI, or, in the north AEH. When a system of interrelated direction courses (Table V) occurs the direction is marked by FI cover and accentuated by DFI margins. Often it obtains that beyond the fringe there is an expanse of FI cover. Where the system denotes a major course divided into parelleled components, each of the latter is covered with FI and is separated from its neighbour component by EI cover.

When the complete physiographic implication in organic terrain is assessed, the approach is not so complicated as the foregoing discussion might suggest. Simplification is achieved by the arbitrary procedure of always interpreting for physiographic features in a certain order. Then it is less likely that any will be missed and the configurations that are most fundamental will not be neglected through inconsistency of approach. To encourage this principle, mineral configurations are sought out first, subsurfacc ice features next, then those that arise in the organic overburden, and finally the water-imposed conditions (i.e. M.I.O.W.).

If consideration is now given to the task of co-ordinating all vegetal cover factors with the physiographic ones, there are difficulties in spite of uniform well reasoned interpretive procedure. This is largely because of the great number of variables involved in analyses and partly owing to a degree of overlap particularly among cover formulae. To facilitate comprehension of the physiographic factor in relation to vegetal cover, Table VI has been prepared. To do this cover formulae were incorporated into their appropriate families which are designated in the table by cover class symbols A to H. Below each of these, the order of prominence of the four physiographic categories is indicated. The histograms in the table show the relative extent to which the categories are important in terms of area of terrain covered and the number of kinds of configurations represented. The physiographic categories have been arranged in the conventional M.I.O.W. order, e.g. with "Mineral" at the top of the histogram series for each family and "Water" at the bottom.

The difficulty is not so profound if the investigator commences his total interpretation of a given major area by selecting in turn major cover formulae and mapping the physiographic features appropriate to them. It will be noticed that for such an area fewer configurations arise than might have been expected. An imaginary case is suggested in Plate I, i-xii. The cover formula examined is FI and the common physiographic circumstances for which it applies are portrayed conventionally in four sets of line-diagrams in the M.I.O.W. sequence. For a natural case it is highly unlikely that twelve physiographic features would occur in a given area. Despite this, some analysis of the cases schematized in Plate I might be helpful if only for summary elucidation of the foregoing discussion.

Mineral conformation pertinent to FI cover is expressed in the first line of diagrams in Plate I. Space does not permit extensive representation of each configuration but, as for all the diagrams on the Plate, enough is shown to express the main aspect of form, including relative elevation cover conditions and, at the front edge of each diagram, the relative depth of peat. In the latter connection the comparison is only very approximate among diagrams but is thought to hold approximately for each figure. It is shown that the three kinds of configuration differ from all others represented. Superficially there is some agreement between figures i and vii, but it will be recalled that the crescentic and other features of the organic configurations are markedly narrow as compared with those of the mineral. Casual inspection also suggests that figure ii might represent the same condition as figure vi. In both cases cover formula FI characterizes the background and is interrupted. More intensive comparison shows however the interruptions shown in figure vi are in the subsurface ice not in the cover and in the mineral sub-layer as they are in figure ii.

Within the first three diagrams one configuration (figure i) is covered by a formula other than FI, another (figure ii) by FI, and the third bears no organic terrain. The first condition is found where there are old beach lines; the second where ponds have become closed or overgrown with

FI cover; and the third where igneous rocks in their folding enclosed wide trenches of organic terrain with FI cover, a condition which is common in northwestern Quebec.

The configurations caused by the ice factor are obscure except for the polygonous condition portrayed in figure iv. The shallow depth of peat shown on the areas contained by the polygonal trenches is not typical south of Fort Churchill in Manitoba. Here the peat accumulates in the areas to a depth of about three feet above the top of the peat in the polygonal boundaries. Relative to these boundaries the fissures (figure v) are narrow and represent a different phenomenon altogether (6).

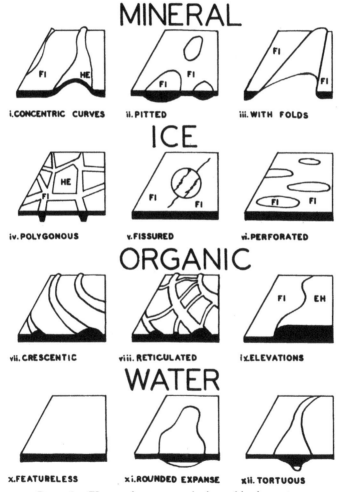

PLATE 1.   FI cover in common physiographic circumstances

A comparison of figures vii, viii and ix suggests that cover formula FI might prevail in the depressions. This is likely to be so for figure vii, but not necessarily for figure viii, where cover FI is sometimes confined to the elevated units and open water occurs among them. When FI is, in fact, found in the depressions, the reticulations bear cover usually of a formula in the B family, such as BFI or BDE. In this case the peat as a whole will be deeper. The shape of the area bearing FI cover in figure ix is distinctive as contrasted with that typical for crescentic and reticulated conditions.

In the last row of diagrams showing water-imposed conditions, the featureless one signifies an even depth of peat. The rounded expanse (figure xi) encloses deep FI covered peat (twenty to thirty feet to water) which usually quakes with applied vibrations. The tortuous condition represented in figure xii contains peat of variable depth, ten to fifteen feet on the average and is also FI covered.

If space permitted, examples based on other cover formulae could be offered. Perhaps even with a single cover formula, enough has been demonstrated to illustrate how organization is reflected in organic terrain on the framework of vegetal and physiographic relationship. The conditions represented in Plate I are real; artificiality obtains only in the subjective arrangement of the diagrams.

Recurrence of cover-physiographic phenomena also reflects a degree of organization contributing to the characterization of organic terrain. Recognition of this encourages interpreters to begin extending their methods to identify the aggregate in the mineral sub-layer. For example, AH cover in association with axial foundation in mineral terrain foundation typically denotes presence of fine to coarse sand. AEH in similar circumstances suggests a foundation of gravel containing varying amounts of silt, and if the cover is AEI the gravel is also suggested but with still more silt, enough to make the gravel "dirty." If HE is associated with polygonal conformation, sandy silt is likely to be at the base of the peat whereas if the same cover co-exists with axial mineral foundation, coarse gravel is likely to occur at the base of the relatively shallow peat beneath it. DFI and DEI on paired sinuous axial foundation (for example, elevated banks of a river draining through organic terrain in the north) indicate presence of gravel with coarse isolated boulders. At the outer edges of these configurations the gravel becomes silty and bears cover formula of AEI. Beyond this, in a featureless areal foundation, there is liable to be a silty clay beneath deep peat bearing EI cover, changing at intervals to clay usually where there is FI cover.

These comparisons, even though they are based on observation rather than on rules, may have useful application; but the object in presenting them is to expose relationship and to enhance the principle that organic terrain demonstrates cosmic associations. This kind of terrain, once formidable to the investigator, is now becoming one of the easiest to interpret. If the mechanical properties of the matter comprising it can be properly appraised, ın accordance with its cosmic constitution, engineering implementation ın the north can proceed with confidence.

## ACKNOWLEDGEMENTS

The author is indebted to the National Research Council and to the Defence Research Board for their financial support of the muskeg research programme. To his students and assistants past and present, and especially to Mr. K. Ashdown for constructing tables and diagrams the writer is also indebted. Gratefully, acknowledgement goes to R. F. Legget, Chairman, Associate Committee of Soil and Snow Mechanics National Research Council for his personal encouragement (since 1945) and interest as applied to the muskeg problem generally.

## REFERENCES

1. ALLINGTON, K. R. (1959). The bogs of central Labrador-Ungava: an examination of their physical characteristics. McGill Sub-arctic research papers, no. 7.
2. FAEGRI, K., and IVERSEN, JOHS (1950). Textbook of modern pollen analysis. Copenhagen: Ejnar Munksgaard.
3. GALKINA, E. A. et al. (1959). Peat bogs of Karelia. Trans. of the Karelian branch of the Academy of Science, U.S.S.R., vol. 15.
4. RADFORTH, N. W. (1952). Suggested classification of muskeg for the engineer. Engineering Journal, 35, no. 11.
5. ——— (1953). The use of plant material in the recognition of northern organic terrain characteristics. Trans. Roy. Soc. Canada, Series III, 47, Sec. V: 53–71.
6. ——— (1954). Palaeobotanical method in the prediction of subsurface ice conditions in northern organic terrain. Trans. Roy. Soc. Canada, Series III, 48, Sec. V: 51–63.
7. ——— (1955). Range of structural variation in organic terrain. Trans. Roy. Soc. Canada, Series III, 49, Sec. V: 59–65.
8. ——— (1955). Organic terrain organization from the air (Altitudes less than 1,000 feet). Handbook No. 1, DR No. 95, Defence Research Board, Dept. of National Defence, Canada.
9. ——— (1956). Muskeg access, with special reference to problems of the petroleum industry. Trans. Can. Min. Metall. Bull., Petroleum and Natural Gas Division, Calgary, vol. LIX, 271–7.
10. ——— (1958). Organic Terrain organization from the air (Altitudes 1,000 to 5,000 feet), Handbook No. 2, DR No. 124, Defence Research Board, Dept. of National Defence, Canada.
11. ——— and L. S. SUGUITAN (1959). Definitive microfossils pertinent to physiographic difference in muskeg. Trans. Roy. Soc. Canada, Series III, 53, Sec. V: 35–41.

# CLAY MINERALOGY OF CANADIAN SOILS*

## S. A. Forman and J. E. Brydon

IN THE STUDY of the clay mineralogy of soils there are two major objectives: (a) a knowledge of the regional distribution of clay minerals and their relation to lithology; and (b) an understanding of the effects of pedo-chemical and other processes on the clay minerals, and vice versa. This paper reviews current information on the clay minerals of Canadian soils and their interrelations and origins.

In spite of the relatively meagre literature on the clay mineral constitution of Canadian soils, certain deductions can be made regarding regional distribution of the minerals. Some of this information is available in papers and reports on soil investigations (3–9; 17; 21; 26; 27; 28; 30; 32; 36; 37) and some is available in a few papers and reports on geological and mineralogical studies of preglacial and glacial deposits (1; 12; 18; 20; 23). A modest amount of information is available in unpublished reports of the Mineralogy Section, Soil Research Institute, Canada Department of Agriculture, Ottawa.

### DISTRIBUTION OF CLAY MINERALS

As a first approximation, the clay minerals of soils in Canada may be said to be mainly illite in the east and montmorillonite in the west. Brydon (3, 4) in a study of selected horizons of soils from 34 localities in the Maritime Provinces, found illite to be the predominant clay mineral, with chlorite, vermiculite, and kaolinite being often present in significant amounts. An appreciable amount of montmorillonite occurred in only one soil, the Truro, which was related to the underlying Triassic Red Sandstone. This agrees with the findings of Allen and Johns (1) who found that the predominant clay mineral in thirteen clays and shales from the Maritime Provinces was illite, significant amounts of chlorite usually being present. In two other clays, the predominant clay mineral was kaolinite. These two clays were considered to be either Late Jurassic or Early Cretaceous, whereas the others were Mississippian, Pennsylvanian, or Quaternary. Brydon and Heystek (6) studied the profiles of six dikeland soils of Nova Scotia and found that illite was the predominant clay mineral throughout.

In the Eastern Townships of Quebec, surface samples of the Greensboro soil, which is thought to be derived from Ordovician limestones and shales,

*Contribution No. 16, Soil Research Institute, Research Branch, Canada Department of Agriculture, Ottawa.

consisted predominantly of illite (4, 7). Eight of thirteen samples of shales and clays in southern Quebec were predominantly illite; three of Precambrian age were predominantly kaolinite; and two were vermiculite with significant amounts of illite (1). Karrow (23), in a study of the Pleistocene geology of the Grondines (Three Rivers) area found the tills, varved clays and marine clays all to be "rock flour," containing illite, chlorite, quartz, feldspar, and amphibole.

The same assemblage of minerals as in the Three Rivers area was found in the marine clays in the Ottawa Valley (1, 4, 9), together with montmorillonite (9) and mixed-layer clays (1, 9). The Silurian and Ordivician shales and the surface soil of the Grenville soil series developed from till in this area and consist of illite with small amounts of chlorite and vermiculite (1, 4).

Of fifteen shales and clays from southern Ontario, thirteen were predominantly illite and two vermiculite (1). Several contained significant amounts of chlorite. This is supported to a large extent in the study conducted by Webber and Shivas (36) on the parent materials of a number of soils in southern Ontario, and one in northern Ontario. In all cases, illite was the predominant clay mineral, montmorillonite being present in several. A surface soil sample of water-laid material from the Lakehead consisted predominantly of an illite-montmorillonite mixed-layer mineral (4).

A study (17, 18) of profiles of ten Manitoba soils on Mankato till showed montmorillonite to be, in general, the predominant clay mineral with illite usually as a major constituent. Other clay minerals were present in minor amounts.

Christiansen (12), in his study of the glaciation of the Swift Current area of Saskatchewan, found montmorillonite to be the predominant clay mineral in tills. He considered the source to be the Bearpaw Formation and, where kaolinite was present in moderate or major amounts, he attributed the source to the Whitemud Formation. Rice, Forman and Patry (28) found that five profiles on lacustrine material from four major soil zones in southern Saskatchewan and Alberta had montmorillonite as the predominant clay mineral. In addition, an investigation of some core samples from the Bearpaw Formation in Saskatchewan (20) suggested a relationship between the montmorillonite found in it and in the soils. Warder and Dion (32) in a study of the clay mineralogy of eight widely separated Saskatchewan soils of different lithology and Great Soil Group, also found that montmorillonite was the major clay mineral, significant amounts of illite being present. Muir (26) found montmorillonite present in a Grey Wooded profile from this area, and Brydon (4) found a Brown surface soil to consist predominantly of an interstratified illite-montmorillonite.

In Alberta, montmorillonite was found to be the predominant mineral in two Grey Wooded soils developed on till (21) and in a Grey Wooded and a Thin Black soil developed on lacustrine materials (28), and this mineral was associated with illite in each case. Pawluk (27), in a study of a podzol derived from a low clay, deltaic sand, found the clay in the C horizon to consist of illite, montmorillonite, and kaolinite.

In British Columbia the preglacial geology was evidently so complicated that, in most cases, each geological deposit is confined to a small area. Consequently, glacial deposits show apparently great variation within short distances. Unpublished records of the Ottawa Institute show that two profiles from the Fraser River Valley had illite as the predominant clay mineral. In a study of the C horizons of three soils (Newton stony marine clay, Rycroft clay and Prince George clay), Clark (private communication) found that illite was the major clay constituent, significant amounts of chlorite being present. The parent material of the Alberni soil series, a Concretionary Brown soil developed from marine clay, was found to be a mixed-layer montmorillonite-chlorite, and the surface soil was mainly a dioctahedral chlorite (8). Theisen et al. (30) found chlorite in the surface layer of three soils on marine material but of differing Great Soil Groups. A mineralogical study (37) of a chronosequence of soils on alluvial deposits in the Northwest Territories showed that the major clay mineral constituent was illite.

In general, therefore, there is substantial evidence that illite is the predominant clay mineral in eastern soils, montmorillonite predominating in the soils of the prairies. British Columbia appears to be too complicated to be described in a general way and there is insufficient information to generalize regarding the Yukon and the Northwest Territories.

## Origins of Clay Minerals

It is generally considered that the composition of most glacial deposits reflects the composition of the nearby preglacial rock formations (19). Consequently the composition of the parent material of a soil, including the clay mineralogy, should be related to the composition of neighbouring rocks. This oversimplified picture is sometimes contradicted by the fact that glaciers are known to have carried material for hundreds of miles. Added to this are many complexities due to (a) successive periods of glaciation; (b) type of glaciation process responsible for the particular deposit; and (c) the nature of the preglacial rock formation, all of which make interpretation difficult. Nevertheless, it has been accepted by students of glacial geology that generally drift has been deposited at a short distance from its origin.

Water-laid materials may or may not reflect the composition of neighbouring rocks, depending on the distance the materials may have been transported by glacial or more recent lakes and rivers. In lacustrine deposits it may be expected that the clay minerals, which may remain suspended indefinitely in fresh water, would have originated in all parts of a given watershed. It must be concluded, therefore, that the clay minerals of lacustrine soils and sediments will be of mixed origin and will not reflect a nearby rock formation unless the entire watershed lay within it. According to Weaver (33–35), the clay minerals in sedimentary rocks (a) are detrital in origin, (b) reflect the character of the source material, and (c) are little changed in their depositional environment. It may be presumed, therefore, that the clays in any unconsolidated deposit due to glacial activity, whether till, glacio-fluvial, marine or lacustrine, will be similar to the source materials. Furthermore, since most of Canada is covered by some form of glacial deposit, these generalizations may be considered to apply to the whole country.

From a practical point of view, the origins of the clay minerals in any deposit can be elucidated only if their specific relationships, based on appropriate factors, can be established. The parent materials of the lacustrine soils of southern Saskatchewan (28) and the tills in that area (12) were related to the Bearpaw shales (20). Christiansen (12) also related the moderate kaolinite content of the till to the kaolinite-rich Whitemud formation. In contrast to this, the marine deposits of Champlain age in the Ottawa-St. Lawrence valleys (1, 4, 9, 23) consist of a variety of minerals in the clay fraction. Several of these minerals, however, that occur even in the fraction less than 0.2 micron in size (9)—the feldspars, quartz and amphiboles—are non-clay minerals, and are thought to be due to the grinding of the nearby Precambrian granitic rocks by glaciers (23).

The above examples show that soils may inherit any or all of the clay minerals and primary minerals of clay size depending on the nature of the source rock. Extreme caution must be used in assigning a given soil to a weathering stage unless a depth function such as that of Jackson et al. (22) is used.

WEATHERING OF CLAY MINERALS

Information on soils of eastern Canada and of the northeastern United States show that chlorite and mica may both be changed to expanding layer silicates. In podzols from Nova Scotia, New Brunswick and Quebec, the illite was found to alter to a mixed-layer illite-montmorillonite mineral (5). This may be compared with the mica weathering products (a) beidellite (2), (b) vermiculite (24, 31), (c) dioctahedral vermiculite (29), and (d) mixed-layer illite-montmorillonite (14, 15, 16) found in other soils.

These changes are not unexpected since it has been demonstrated that potassium can be removed from mica by chemical means (10, 25) or by plant growth (10), the product being vermiculite. It is not known whether, in having potassium stripped out of the interlayer position, certain fundamental changes occur in the tetrahedral positions, such that the ratio of $Al:Si$ is brought closer to that of the known expanding minerals. There is evidence, however, that the product retains at least some of the octahedral characteristics of the original (29, 31). The micas have a larger theoretical exchange capacity than vermiculites and montmorillonites on the basis of their tetrahedral charge. The reverse process, potassium fixation, has been demonstrated in the laboratory (10) and in the field (13). Weaver (33–35) states that the most common process acting on clay minerals in marine environments is cation adsorption. He considers that the modifications resulting from this are secondary and that the basic clay structure remains unaltered.

The alteration of chlorite to expanding-layer silicates has been observed in soils of northeastern United States (14, 15, 16) and in podzols of eastern Canada (5). This reaction has been also demonstrated in the laboratory (6). With orthochlorites, it apparently involves the dissolution of the interlayer hydroxide of magnesium or ferrous iron, leaving a trioctahedral expanding-layer silicate. The reverse process "chloritization," which involves the formation of interlayer hydroxides between the sheets of expanding layer silicates, has been successfully demonstrated in the laboratory with magnesium, nickel, cobalt and aluminum (11). Chloritization involving aluminum has been found to occur in acid soils of British Columbia (8) and Alberta (27).

There is no direct evidence that destruction of clay minerals in Canadian soils is significant in extent, degree, or amount. Undoubtedly it does occur. The lack of evidence may be due to the moderate intensity of the process, to limitations in analytical and sampling techniques, or to the difficulty in carrying out extensive investigation in a country of Canada's size. Rice *et al.* (28) found some evidence of clay destruction in the A horizons of the Brown and Dark Brown soils of southern Saskatchewan and Alberta. In addition, montmorillonites of different types were found in the A and C horizons of a Grey Wooded soil (26) and recrystallization of the clay material from its component oxides was suggested.

Together with the lack of definite evidence of clay mineral destruction, there is no direct evidence of the formation of clay minerals in Canadian soils except as the alteration products of the inherited clay minerals. Kaolinite has not been found to be a major clay mineral in the Canadian soils; where present it is probably inherited. There is no extensive accumulation of iron or aluminum oxides although small amounts undoubtedly occur in most soils, particularly the B horizon of podzols.

Evidently, therefore, the principal weathering process known to be involved in Canadian soils, from the arid Brown soils to the heavily leached podzols, is an intense cation exchange reaction whereby mica and chlorite alter to expanding-layer silicates and, in certain cases, vice versa.

## REFERENCES

1. ALLEN, V. T., and JOHNS, W. D. (1960). Clays and clay minerals of New England and Eastern Canada. Bull. Geol. Soc. Amer. *71*: 75–86.
2. BRAY, R. H. (1937). Chemical and physical changes in soil colloids with advancing development in Illinois soils. Soil Sci. *43*: 1–14.
3. BRYDON, J. E. (1958). Mineralogical analysis of the soils of the Maritime Provinces. Can. J. Soil Sci. *38*: 155–60.

4. ——— (1958). The chemistry of soil potassium: mineralogy. Unpublished report.

5. ———. Mineralogy of three podzol profiles from Eastern Canada. In preparation.

6. ——— and HEYSTEK, H. (1958). A mineralogical and chemical study of the dikeland soils of Nova Scotia. Can. J. Soil Sci. *38*: 171–86.

7. ——— and SOWDEN, F. J. (1959). A study of the clay-humus complexes of a chernozemic and a podzol soil. Can. J. Soil Sci. *39*: 136–43.

8. ——— CLARK, J. S., and OSBORNE, VINCENT (1961). Dioctahedral chlorite. Can. Mineral., *6*, no. 5. In press.

9. ——— and PATRY, L. M. Mineralogy of a Rideau clay soil profile and some Champlain Sea sediments. Submitted for publication, Can. J. Soil Sci.

10. CAILLÈRE, S., and HÉNIN, S. (1949). Transformation of montmorillonite minerals into 10A micas. Min. Mag. *28*: 606–11.

11. ——— and ——— (1949). Experimental formation of chlorites from montmorillonite. Min. Mag. *28*: 612–20.

12. CHRISTIANSEN, E. A. (1959). Glacial geology of the Swift Current area, Saskatchewan. Dept. of Min., Res. Rept. no. 32, Saskatchewan.

13. DE TURK, E. E., WOOD, L. K., and BRAY, R. H. (1943). Potash fixation in corn belt soils. Soil Sci. *55*: 1–12.

14. DROSTE, J. B. (1956). Alteration of clay minerals by weathering in Wisconsin tills. Bull. Geol. Soc. Amer. *67*: 911–8.

15. ——— and DOEHLER, R. W. (1957). Clay mineral composition of calcareous till in northwestern Pennsylvania. Trans. Ill. State Acad. Sci. *50*: 194–8.

16. ——— and THARIN, J. C. (1958). Alteration of clay minerals in Illinoian till. Bull. Geol. Soc. Amer. *69*: 61–8.

17. EHRLICH, W. A., RICE, H. M., and ELLIS, J. G. (1955). Influence of the composition of parent materials on soil formation in Manitoba. Can. J. Agric. Sci. *35*: 407–21.

18. ——— and ——— (1955). Post-glacial weathering of Mankato till in Manitoba. J. Geol. *63*: 527–37.

19. FLINT, R. F. (1948). Glacial geology and the Pleistocene epoch. New York: John Wiley and Sons.

20. FORMAN, S. A., and RICE, H. M. (1959). A mineralogical study of some core samples from the Bearpaw formation. Can. J. Soil Sci. *39*: 178–84.

21. HORTIE, H. J., and RICE, H. M. The mineralogy of Cooking Lake and of Breton soils. In preparation.

22. JACKSON, M. L., TYLER, S. A., WILLIS, A. L., BOURBEAU, G. A., and PENNINGTON, R. P. (1948). Weathering sequence of clay-size minerals in soils and sediments. I. Fundamental generalizations. J. Phys. Coll. Chem. *52*: 1237–60.

23. KARROW, P. F. (1957). Pleistocene geology of Grondines map-area, Quebec. Ph.D. thesis, University of Illinois.

24. KREBS, R. D., and TEDROW, J. C. F. (1957). Genesis of three soils derived from Wisconsin till in New Jersey. Soil Sci. *83*: 207–18.

25. MORTLAND, M. M., LAWTON, K., and UEHERA, G. (1956). Alteration of biotite to vermiculite by plant growth. Soil Sci. *82*: 477–81.

26. MUIR, A. (1952). Report of Rothamsted Experiment Station, p. 54.

27. PAWLUK, S. (1960). Some podzol soils of Alberta. Can. J. Soil Sci. *40*: 1–14.

28. RICE, H. M., FORMAN, S. A., and PATRY, L. M. (1959). A study of some profiles from major soil zones in Saskatchewan and Alberta. Can. J. Soil Sci. *39*: 165–77.

29. RICH, C. I. (1958). Muscovite weathering in a soil developed in the Virginia Piedmont. Clays and clay minerals. N.A.S.-N.R.C. Publ. 566, 203–12.

30. THEISEN, A. A., WEBSTER, G. R., and HARWARD, M. E. (1959). The occurrence of chlorite and vermiculite in the clay fraction of three British Columbia soils. Can. J. Soil Sci. *39*: 244–51.

31. WALKER, G. F. (1949). The decomposition of biotite in the soil. Min. Mag. *28*: 693–703.

32. WARDER, F. G., and DION, H. G. (1952). The nature of the clay minerals in Saskatchewan soils. Sci. Agric. *32*: 535–47.

33. WEAVER, C. E. (1958). Geologic interpretation of argillaceous sediments. Part I. Origin and significance of clay minerals in sedimentary rocks. Bull. Amer. Assoc. Pet. Geol. *42*: 254–71.

34. ———— (1959). A discussion of the origin of clay minerals in sedimentary rocks. Proc. Fifth National Conf. Clays and Clay Minerals, N.A.S.-N.R.C. Publication 566, 159–73.

35. ———— (1959). The clay petrology of sediments. Proc. Sixth National Conf. Clays and Clay Minerals, N.A.S.-N.R.C. Earth Science Series, Monograph no. 2, 154–87.

36. WEBBER, L. R., and SHIVAS, J. A. (1953). The identification of clay minerals in some Ontario soils: 1. Parent materials. Proc. Soil Sci. Soc. Amer. *17*: 96–9.

37. WRIGHT, J. R., LEAHEY, A., and RICE, H. M. (1959). Chemical morphological and mineralogical characteristics of a chronosequence of soils on alluvial deposits in the Northwest Territories. Can. J. Soil Sci. *39*: 32–43.

# THE SOILS OF CANADA FROM A PEDOLOGICAL VIEWPOINT

## A. Leahey

THE WORD "SOIL" has been defined in many ways and it is being used with different meanings. The simplest definition from an agricultural viewpoint is that soil is the natural medium for the growth of land plants. From a pedological viewpoint, however, a more adequate definition is that soil is the collection of natural bodies on the earth's surface supporting or capable of supporting plants. These natural bodies may be divided into mineral or organic soils but in this paper attention will be focused almost entirely on mineral soils.

Soils have length, breadth and depth; that is, they are three dimensional bodies. Their upper limit is usually air but may be water. At the lateral margins, each soil may end abruptly but is more likely to grade into other soils, bare rock or deep water. The lower limit is the most difficult to define. One concept is that soil has a thickness determined by the depth of rooting plants. Another concept is that soil depth should be restricted to the thickness of the solum. While there are theoretical justifications for these concepts, neither is satisfactory for the mapping and classification of soils. In Canada the lower limit is usually considered for classification purposes to be about four feet.

Soils have morphological features that have developed during and by the process of soil formation. These morphological features are expressed by soil horizons which have been designated by the letters A, B and C. The A and B horizons are a reflection of the genetic forces operating on the mineral parent material; together they constitute the solum. No simple definition of the A and B horizons is possible since there are many different kinds of both horizons even in one soil. The A horizons contain most of the organic matter deposited by plants and most of the micro-organisms; they have been subject to the greatest amount of weathering and leaching of the mineral matter. The B horizons, lying immediately below the A horizons, contain most of the material leached from the A horizons and have other changes brought about by weathering and pedogenesis. The C horizons represent the relatively unweathered underlying geological deposits and are usually the parent materials of the overlying sola. The C horizons have been modified to some small extent by biological activity and by receiving some mineral material, usually easily soluble salts, leached from the A and B horizons. Soils vary widely in the nature of all these principal horizons and in the degree of expression and thickness of the A and B horizons. The depth to which soils

147

are examined in the field nearly always includes at least the upper part of the C horizon. Thus, soil descriptions usually contain a description of the upper portion of the geological deposit or parent material of the soil.

Knowledge of Canadian soils comes largely from the work of the soil surveys which are carried out co-operatively by the Canadian Department of Agriculture, the colleges of agriculture, the provincial departments of agriculture and provincial research councils. These surveys are charged with the task of mapping, describing and classifying the soils in the agricultural and potentially agricultural regions of Canada. Since these surveys are carried out primarily for agricultural purposes, interpretation of the soils information obtained is, naturally, from an agricultural viewpoint. Experience has shown, however, that the basic information provided by the soil surveys can be used, if properly interpreted, for many other purposes.

The literature on Canadian soils is too voluminous for even brief review here. Three sources of information which are perhaps of most direct concern to geologists and engineers may usefully be mentioned. These are: (1) The soil survey reports and maps of the reconnaissance surveys: to date maps covering about 180,000,000 acres have been published so that soil information is available for most of our settled areas and for a considerable portion of the fringe areas. (2) The March-April 1960 issue of the *Agricultural Institute Review*: this issue, entitled "A Look at Canadian Soils," is devoted to papers that present a broad picture of the soils in the various physiographic regions of Canada and their classification. (3) A paper by H. C. Moss entitled "Modern Soil Science (Pedology) in Relation to Geological and Allied Science" in the *Transactions of the Royal Society of Canada, Volume LIII; Series III: June 1959.*

Soil mapping and classification is not a static field of investigation. This natural body known to the pedologist as soil is a mysterious body in many ways. Concepts are developed as to its origin, the significance of its morphological, chemical, physical and biological properties and its behaviour under treatments imposed by man. These concepts are limited by the sphere of knowledge at any one time and consequently, as the body of pertinent knowledge increases, concepts change. This gradual evolution of ideas affects the whole field of work generally included in the term soil survey. Thus the early soil survey publications differ markedly from those issued today as a result of evolving concepts of soil genesis and classification, accompanied by changes in terminology.

## SOIL FORMATION

The nature and distribution of Canadian soils is the result of many soil-forming factors. Some consideration of these factors is presented here as background information and to illustrate the pedological viewpoint on soils. The main natural soil forming factors are the mineral parent material, the climate, vegetation, drainage conditions and time. Soil genesis can be viewed as consisting of two steps; (a) the accumulation of parent materials and (b) the differentiation of horizons in the profile. Climate and vegetation may be considered as the main active forces in soil formation, but their effect is conditioned by the nature of the mineral material, the moisture status, and the length of time they have been at work.

Horizon differentiation in the mineral parent material is brought about by the weathering of minerals, the accumulation and assimilation of organic matter, removal out of and transfers and transformations within the soil system, and the development of structure. While the imprint of the original parent material is clearly evident in soils, the processes of soil development have in most cases produced great changes in both chemical and physical properties. Thus the various horizons of the soil may have markedly different properties from each other and from the parent material.

*Parent materials*

The geological nature of soils certainly receives as much attention today as it did a half century ago when scientists thought of soil as disintegrated rock mixed with some decaying organic matter. The significance of the geological nature of soils is clearly shown by the system used in classifying the individual soils in soil survey reports. While the pedologist is mainly concerned with the geology of the upper few feet of the regolith, sometimes he must examine deeper layers and even the bedrock in seeking answers to his problems.

A large number of factors that affect the properties, classification, use and management of soils are directly related to the geology of the parent material. Some of these are particle size, degree of sorting, chemical composition and nature of the soil minerals, kind and quantity of salts, permeability and drainage, land form and topography. Hence the pedologist must be concerned with both the mineralogical nature and the mode of accumulation of the regolith. He is also concerned with the interrelations of the parent material with other soil forming factors.

*Climate*

In Canada, climate has both direct and indirect effects on soil formation. The indirect effect of climate which is exerted through its influence on vegetation is, with one exception, more conspicuous than its direct effect. The exception is the occurrence of permafrost which severely restricts the operation of normal soil forming processes. The direct effect of climate has not been the subject of adequate research in this country. Certainly it would appear that the depth to which Canadian soils are leached is a function of the effective precipitation. It would also appear that climate is responsible for the geographic distribution of Acid Brown Wooded (Brown Podzolic) soils in relation to Podzols, the occurrence of secondary podzolization in the Grey Wooded and Grey Brown Podzolic soils, and for the unique character of a group of soils on the west coast.

*Vegetation*

The three great vegetative regions of Canada are the grasslands in the western provinces, the forest region which covers vast areas from coast to coast and which extends well into the permafrost regions, and the tundra. The heath covered parts of Newfoundland can be considered as a fourth vegetative region. Outside permafrost regions, these kinds of vegetation produce major differences in soils, differences which greatly affect soil classification. The effect of trees rather than tundra vegetation on soils with permafrost is however rather obscure, since present scanty knowledge indicates that the characteristics of the wooded soils with permafrost are similar to those of the tundra.

*Drainage*

The full effect of climate and vegetation on soil formation can only be clearly expressed on reasonably permeable parent material which can absorb the precipitation it receives without becoming water-logged. If the topography is rolling, run-off may occur and collect in the low places. In such instances, the upper part of the slopes may be more arid and the low places much wetter than is normal for the region. This results in thin sola on the drier slopes and soils developed under wet conditions in the low places. In humid regions, if the topography is level and the parent materials only slowly permeable, soils also develop under wet conditions which result in the formation of quite different soils from the well-drained soils of the regions.

*Time*

Soils, being dynamic bodies, progress from youth through maturity to old age. Although these stages of soil development are not entirely related to the length of time during which the genetic forces have been at work, yet it does take time for such forces to show their effect on soils. In terms of years none of the Canadian soils is very old, all of them dating from the last ice age and some of them being much younger. This youthfulness in point of years no doubt accounts for the fact that the solum of Canadian soils is thinner and less weathered than in many other countries of the world. Despite their comparative youth, many have reached a mature state of development.

## Soil Classification

Soils may be classified in many ways depending on the purpose of the grouping. The pedologist must however base his classification on those features that he can observe in the soil profiles, including the mineral parent material. Furthermore, apart from the basic units of soil classification, he must be selective in the features he uses for classification purposes in order that he may be able to group soils at progressively higher levels of abstraction. In Canada, morphological features which are largely a reflection of the effects of climate, vegetation, local moisture relations and age have been selected for these higher groupings, although parent material does play a part in many instances. In other words, although soil classification in Canada is based on morphological features, concepts of soil genesis affect the selection of criteria used for the higher groupings.

Since the soil surveyor is concerned with the mapping of soils at various scales and at different levels of abstraction, as well as the grouping of soils on the basis of common or similar profile characteristics, two kinds of classification systems are used in Canada. One of these may be referred to as a mapping or geographic classification and the other a taxonomic classification. Since soils which are closely associated geographically may differ markedly in their morphological characteristics, it has not been found possible to devise one system for both purposes.

## NATIONAL TAXONOMIC SYSTEM OF SOIL CLASSIFICATION

Although taxonomic groupings were in use in some provinces as far back as 1927, it was not until 1955 that the National Soil Survey Committee first proposed a national system. This proposal has been studied across Canada since then. In February 1960 the soil survey organizations accepted for official usage a revised system based on the principles established in 1955. This new national system does not as yet appear in any of the soil survey publications but it will be used in the future. Since most of the terminology and definitions now in use have been retained in the new system, those familiar with the classification followed in modern soil survey reports should have no particular difficulty in understanding it.

The taxonomic system places soils in categories at different levels of generalization on the basis of their internal characteristics. Six categories are used. These are:

Category 1: *The soil type*: this is a sub-division of the soil series based on the texture of the surface soil; hence its retention in the classification system as the lowest category is perhaps more a matter of traditional use than logic.

Category 2: *The soil series*: this is the basic unit of classification in the system since it is the natural body in the definition of soil from a pedologist's viewpoint; the series in some soil survey reports is named a "member" or an "associate."

Category 3: *The soil family*: this is a grouping of series belonging to the same subgroup on the basis of some important characteristics in the parent material; this category has not as yet been widely used in Canada.

Category 4: *The subgroup*: the subgroup is analogous with the types of profiles long recognized in Canada.

Category 5: *The great group*: this is a grouping of the subgroups on the basis of profile similarities.

Category 6: *The order*: the great groups are organized into orders on the basis of major profile similarities.

This taxonomic system will permit the study of soils at various levels of abstraction. This is shown by the fact that the large number of soil series identified in Canada to date—some 1,800—can be placed in about 100 subgroups, 23 great groups and 6 orders.

A brief description here of the six orders and the names of the great groups may serve to show the principles on which this system of classification is based.

### Chernozemic Order

Most of the soils in the grassland and park areas of western Canada are placed in this order. These soils have dark-coloured mineral-organic surface horizons and brownish, usually prismatic, subsurface and usually non-saline horizons lying on calcareous parent material. They are well saturated with bases. The sola are well drained and free of soluble salts. The order includes four great groups, the Brown, Dark Brown, Black and Dark Grey. The division into great groups is based largely on the colour of the surface soil which is a reflection of the climate and vegetation under which these soils have developed.

### Solonetzic Order

The soils in this order have developed from saline parent material or under the influence of saline waters. They occur dominantly in the grassland areas of western Canada but extend into the forested regions. The dominant influence in the formation of the solonetzic soils has been the parent material. The chief characteristic of this order is the presence of a tough, finely textured B horizon which usually breaks into column-like structures or blocky aggregates that have surface coatings of organic matter. Three great groups have been recognized; the Solonetz, the Solodized Solonetz and the Solod.

### Podzolic Order

Most of the well-drained soils in the forested region south of the permafrost belong to this order. They are characterized by an impoverished grey layer near or at the surface of the mineral soil which is underlain by a darker subsurface horizon enriched in clay or organic matter and sesquioxides. Four great groups are placed in this order: the Grey Wooded, the Grey-Brown Podzolic, the Podzols and the Humic Podzols. The Grey Wooded and the Grey-Brown Podzolic soils have been formed from calcareous parent material whereas the Podzols and Humic Podzols have

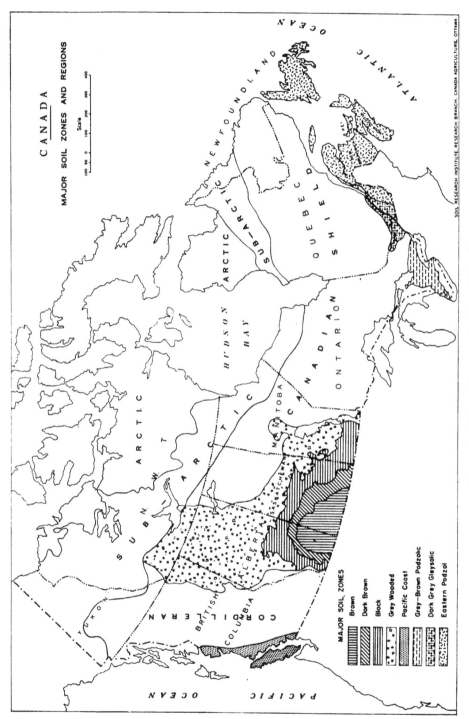

Figure 1. Major soil zones and regions of Canada

developed for the most part on acidic parent materials. In both the Grey Wooded and Grey-Brown Podzolic soils, eluviation has been mainly restricted to movement of clay, while organic matter and sesquioxides have been moved downwards in the Podzols and Humic Podzols.

## Brunisolic Order

For reasons associated with climate, with parent material or with age, a considerable number of well-drained soils in the forest region do not have the leached grey horizon or the enriched subsurface horizon of the Podzolic soils. The sola of these soils are dominantly brown and hence they are referred to as "Brunisolic soils." Five great groups have been recognized: the Brown Forest and the Brown Wooded on calcareous parent material with high base status; the Acid Brown Forest and Acid Brown Wooded or Brown Podzolic on non-calcareous parent materials with low-base status; the Concretionary Browns of the west coast which resemble the Acid Brown Wooded soils but contain magnetic ferruginous concretions. These great groups are considered to be in a youthful state of development.

## Regosolic Order

A number of well-drained soils occur in Canada which lack any noticeable horizon development except, in some cases, a mineral-organic layer at the surface. Essentially these soils are composed of only slightly modified mineral parent material and this fact is indicated by the name Regosolic. There is little basis for dividing these soils into great groups as there has been little or no genetic development. Two great groups have been established, however, one for soils having no horizon development and one for soils with weak development.

## Gleysolic Order

This order includes the poorly drained soils in which normal processes of development have been restricted. The excessive water and lack of aeration have established reducing conditions which are indicated by dull-coloured and mottled subsoils. Unlike the soils mentioned in previous orders, the Gleysolic soils may differ markedly in their profile characteristics. These have been used in setting up four great groups: the Meadow and Dark Grey Gleysolics which are closely related inasmuch as both have dark coloured mineral-organic surface horizons; the Eluviated Gleysols which have podzolic features; and the Gleysols which are in effect just wet Regosols.

## MAPPING OR GEOGRAPHIC CLASSIFICATION

Although the effect of climate and vegetation on soils may be fairly uniform over broad areas, local differences in parent materials and drainage have resulted in the formation of different soils closely associated geographically. Hence soil series often occur in such small areas that they cannot be used as mapping units except on very large scale maps. It is then often desirable to group soils for generalized soil maps. For these reasons, the mapping units may range from a subdivision of a series to groups of series, depending on the scale of the map and the complexity of the soil pattern.

Three terms are commonly used in Canada for the geographic grouping of soils: (1) The catena: This term is applied to the range of soils produced under different moisture relationships on the same parent material within a common climatic and vegetative region. (2) Soil complex: This term is used to designate mixed areas of soils derived from different parent materials. (3) Soil zone: This term is used for a broad geographic grouping of soils on the dominant occurrence of one great group. The boundaries and extent of the zone are largely determined by soil characteristics which can be correlated with climate and natural vegetation. Hence the zone coincides closely with major climatic and vegetative regions. From soil zone maps which have been published, it is obvious, however, that there is also a close relation between the major physiographic regions and soil zones. Soil zone boundaries do not cross major physiographic boundaries, although several soil zones may occur within one major physiographic division.

Pedologists as yet do not know enough about all the unsettled parts of Canada to divide the entire country into soil zones. Those parts not covered by soil zones may be referred to as soil regions, based on the physiography and such knowledge as is available about the soils. Most of the Canadian Shield, most of the Cordillera and the regions north of the southern limits of permafrost are referred to as soil regions.

The largest-scale map published which shows the soil zones and regions in Canada is the generalized soil map in the Atlas of Canada on a scale of 1 to 10,000,000. More detailed information can be found in publications issued by the different soil survey organizations in Canada. In the sketch map in this paper (Figure 1), eight major soil zones and three regions are shown. Chernozemic soils are dominant in three zones, Podzolic soils in three; one zone has Brunisolic soils dominant and the remaining zone Gleysolic soils.

The Chernozemic soil zones cover the grassland soils in the Prairie Provinces. The Brown soil zone lies in the most arid section, the Black soil zone in the most humid section with the Dark Brown soil zone lying between. Solonetzic soils occupy large acreages in these three zones and are in fact the dominant soils in many places. Meadow soils are also of common occurrence on the wetter sites.

The three zones in which Podzolic soils are dominant are the Grey Wooded zone which covers the forested parts of the Great Plains south of the permafrost line, the Grey-Brown Podzolic zone in the Great Lakes region in southern Ontario and the Eastern Podzol zone which coincides with the Appalachian region. The predominant kind of soil in each of these zones is, of course, the great group after which they have been named. Locally, however, other soils may be dominant. In the Grey Wooded zone organic soils, commonly referred to as muskeg, cover large acreages. Other soils of significance in various parts of the zone are Gleysolic and Regosolic soils,

Podzols and Black soils. The Grey-Brown Podzolic zone has a common pattern of Grey-Brown Podzolic soils on the well and imperfectly drained sites and Dark Grey Gleysolic soils on the poorly drained sites. Other soils of importance in some areas are Podzol, Brown Forest, Regosol and organic soils. In the Eastern Podzol zone, the main soil pattern is formed by Podzols on the better drained sites and Gleysols or Eluviated Gleys in the poorly drained sites. Humic Podzols are of widespread occurrence in Newfoundland and occur to some extent in Nova Scotia and New Brunswick. Organic soils also occupy considerable acreages, particularly in Newfoundland.

The St. Lawrence Lowland is a zone in which Dark-Grey Gleysolic soils are dominant, especially on the wide-spread fine-textured deposits. Other mineral soils which occur on moderately to well-drained sites are the Podzol, Brown Forest and, to a lesser extent, Grey-Brown Podzolic, Grey Wooded and Acid Brown Wooded. Organic soils are of common occurrence in some localities.

The West Coast, which is characterized by mild wet winters and cool dry summers, is a zone in which Brunisolic soils (Concretionary Brown or Acid Brown Wooded or, to a lesser extent, Acid Brown Forest Soils) are dominant. Regosols together with organic soils are of particular importance on alluvial flood plains and deltas.

Surveys carried out to date in the areas designated as soil regions indicate that it will be possible to subdivide regions into zones when the pedologists are able to make enough observations. Many parts of these regions are exceedingly difficult to explore and it will no doubt take years to collect the necessary information. Tentative zones for parts of these regions have been shown on the atlas maps for Canada and for British Columbia. In this paper, however, these subdivisions will not be discussed.

The Canadian Shield soil region is, on the whole and as far as is known at present, dominated by Podzols interspersed with organic soils and bare rock. In its southern fringe however the Acid Brown Wooded great soil group is the main type. Several large bodies of calcareous deposits in lacustrine basins occur in this region. On these, Grey Wooded soils associated with organic soils and to a lesser extent with Gleysolic soils form the soil pattern.

The Cordillera is a very complex region from the viewpoint of soils, due to the extreme variations in soil-forming factors which occur within short distances. Both horizontal and vertical zoning occurs. In the drier sections Chernozemic, Brunisolic (especially Brown Wooded), and Grey Wooded soils are dominant, while in the wetter and cooler areas Podzols and Alpine soils occur. There is also, of course, a large amount of rock and rock rubble. Organic soils are of small extent, no doubt on account of the topography.

That part of Canada in which the subsoils are permanently frozen has been divided into two regions on the basis of vegetative cover, namely the Subarctic region with forest vegetation and the Arctic with tundra vegetation. The pattern of mineral soil development in both regions is similar, in that profile development has been hindered by the presence of permafrost.

There has been a little more development in the soils of the Subarctic than in those of the Arctic. Although only a few of these soils have been examined, it would seem that the mineral soils could be classified into the Brunisolic, Gleysolic and Regosolic orders, depending on their drainage and degree of development. Perhaps the greatest contrast between these two regions lies in the fact that organic soils are much more prevalent in the Subarctic Region. The landscapes and nature of the surficial deposits of both regions of course vary greatly between the major physiographic regions.

# CHARACTERISTICS AND GENESIS OF PODZOL SOILS*

## P. C. Stobbe

PODZOL SOILS are natural dynamic bodies, on the surface of the earth, that constitute one of the major great soil groups of the world. They have developed under climatic and biologic conditions that resulted in the accumulation of an organic surface layer and in the formation of acid decomposition products of organic matter. Their formation and their characteristics may be greatly influenced by the nature of the mineral parent materials, but they occur on a wide variety of land forms and on parent materials that differ considerably in geologic origin and in lithology. Relief and drainage may also greatly influence the formation of these soils since they affect the biologic conditions and the percolation of the soil solution through the soil. The time required for the formation of Podzol soils may vary from several decades to thousands of years, depending on the intensity of the factors that control the soil environment.

The Podzol group of soils belongs to a larger order of Podzolic soils that is distinguished by having a light-coloured eluvial horizon at or near the surface (Ae horizon) and an underlying darker accumulation (B) horizon containing some of the products that have been leached from above. In the Podzol group the accumulation products consist mainly of organic matter and sesquioxides; in the other Podzolic groups clay is the main accumulation product.

### CHARACTERISTICS OF THE PODZOL GREAT GROUP

Morphologically the profile of a Podzol soil may be characterized as follows: Under natural conditions the surface consists of an organic layer (O horizon) that may be subdivided into three subhorizons: Undecomposed raw litter (L), fermenting semi-decomposed brown organic matter (F), and humified dark organic matter (H). The organic layer is underlain by a mineral light-coloured eluviated horizon (Ae) from which most of the exchangeable bases and some of the sesquioxides have been lost. This horizon is generally structureless in the coarse sandy soils and platy in the fine clayey soils. It is in turn underlain by a darker brownish or reddish accumulation (B) horizon, in which organic matter and sesquioxides are the main accumulation products. The structure of this horizon may vary from soft,

*Contribution No. 17, Soil Research Institute, Research Branch, Canada Department of Agriculture, Ottawa, Ontario.

friable granules to a hard, cemented, structureless mass. The B horizon gradually fades, or breaks more or less abruptly, into the mineral parent material (C horizon), which has not been changed appreciably by pedogenic processes. The parent material may be friable or compact; it may range in texture from sand to clay; and it may vary greatly in its geologic origin and mineral composition. In general, however, it is non-calcareous, that is, it has been derived from carbonate-free formations or the carbonates have been leached out by weathering processes.

Chemically, the Podzol soils are acid throughout their profile. The reactions of the surface horizons may vary in pH from 3.5 to about 6.0 and they increase slightly with depth. The Ae horizon is somewhat enriched in silica and the B horizon in organic matter and sesquioxides. The cation exchange capacity increases from the Ae to the B horizon, and the degree of base saturation generally decreases from the Ae to the B.

### CHARACTERISTICS OF THE SUBGROUPS

From the above it is evident that the Podzol soils may differ considerably in their morphological and chemical characteristics, many of which are associated with their degree of development. In many Podzols, referred to as Minimal Podzols, development of the entire profile is very weak. In these soils either the Ae is very thin and can easily be destroyed or the accumulation in the B horizon is not very marked. Such profiles resemble morphologically and chemically those of some of the Brunisolic groups of soils in which there is no significant translocation or accumulation in the profile.

The Humic Podzols are confined to cool, moist areas, such as the Atlantic coastal regions, high plateaux, and poorly drained depressional areas in which heath or forest with a heath type of undercover form the dominant vegetation. These soils have a thick organic layer on the surface and contain large accumulations of organic matter (10 per cent to 20 per cent) in the B horizon that impart a characteristic dark reddish-brown to black colour to this layer. The humus-rich B horizon in some of the Humic Podzols also contains considerable amounts of free iron, or it may be underlain by an iron-enriched subhorizon. In other subtypes of the Humic Podzol, free iron may be lacking in the B horizon because of iron-deficient mineral material or of removal of iron from the profile under strong reducing conditions in very poorly drained soils. The Humic Podzols are strongly acid. The cation exchange capacity of the humus-rich B horizon is very high (30 to 40 me./100 gm.), but the degree of base saturation is very low. The B horizon may be friable or strongly cemented and impervious, thus restricting moisture percolation and causing the soil solution to move laterally.

In most of the Canadian Podzol soils, the increase in organic matter from the Ae to the B horizon is considerably less than in the Humic Podzol and this is generally associated with a less dense vegetative cover. The organic matter content of the B horizon may vary from about one per cent to ten

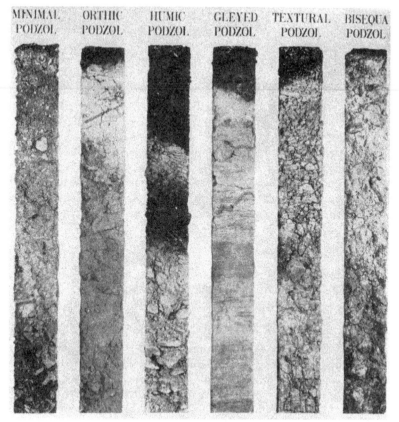

**FIGURE 1.** Profiles showing the development of different kinds of Podzol soils

per cent and is generally highest in the upper part immediately below the Ae horizon. The accumulation of free iron may vary from a fraction of one per cent to more than three per cent. This variation is not directly proportional to the organic matter content but depends to a considerable degree on the composition and ease of weathering of the mineral material. The cation exchange capacity of these soils depends to a large degree on the organic matter content but it is also affected, to a much greater degree than in the Humic Podzols, by the mechanical composition of the mineral parent materials. It may vary from a low of 1 or 2 milliequivalents per 100 gm. in very sandy soils low in organic matter, to more than 40 milliequivalents per 100 gm. in fine-textured soils high in organic matter. The degree of base-saturation of most Podzols is less than fifty per cent, but it may vary considerably from this figure depending on the nature of the mineral material and the amount of organic matter accumulated in the B horizon.

In the typical, or Orthic, Canadian Podzol soils the B horizon is friable and permits relatively free movement of moisture through the upper part of the solum. It may, however, be underlain by compact, only slightly weathered, mineral materials that may appreciably restrict the downward movement of soil moisture.

In the Ortstein Podzol soils, the B horizon is irreversibly cemented with organic matter and sesquioxides. This condition may greatly interfere with the moisture movement and with root penetration. There is considerable morphological evidence that this condition is associated with a change from a wet to a drier moisture regime in the soil. It has been suggested, therefore, that cementation is brought about by desiccation and the resultant oxidation.

Many Canadian Podzol soils differ from those described above in that they have additional characteristics that are generally associated with other great groups of soils. Thus, they may be considered as transitional between the Podzols and some of the other great groups.

Most of the imperfectly to poorly drained Podzols, also referred to as Gleyed Podzols, show marked indications of gleying in addition to the regular podzolic features. "Gleying" refers to the reduction of sesquioxides that takes place in wet, saturated soil and results in the formation of dull bluish-grey colours. Such features are characteristic of the subsoils in the Gleysolic great groups of soils. In the Gleyed Podzols, the Ae and B horizons are generally discoloured, and mottled with dull greyish and bright rust-coloured spots, streaks and splotches. These features indicate alternate reducing and oxidizing conditions brought about by fluctuating water-tables. The intensity of mottling and discoloration is not, however, necessarily an indication of the frequency or range of fluctuations, since it also may be greatly influenced by the nature of the parent material and the amount of free iron present in the soil.

It was stated earlier that the major accumulation products in the B horizons of Podzol soils are organic matter and sesquioxides. In the textural Podzols, however, in addition to the sesquioxides and organic matter, considerable amounts of clay have been translocated from the Ae to the B horizon. This clay movement often extends to a considerable depth into the otherwise unweathered parent material. This type of profile development is generally confined to parent materials containing appreciable amounts of clay that is readily dispersed and then carried downward with the gravitational water. Although these soils resemble the Orthic Podzols in their chemical characteristics, they also have some resemblance in their physical characteristics to the Grey Wooded group of soils. They may therefore be considered as a transitional subtype between the two great groups.

The influence of certain parent materials on the development and characteristics of the profiles is also clearly indicated in the Bisequa Podzols. In this case a podzol profile, having normal Ae and B horizons, is underlain by a pedogenic horizon in which clay is the main accumulation product. This horizon is similar in morphology and chemical characteristics to the B horizons found in the Grey-Brown Podzolic and Grey Wooded groups of soils. These soils have formed from mineral materials that were originally rich in carbonates, and free carbonates generally occur immediately below the clay accumulation layer. Weathering and removal of the carbonates causes dispersion of the clay, which then moves downward on removal of lime. The upper Podzol profile has thus developed in the weathered acid material from which the carbonates and some of the clay have been removed in the earlier stages of development. The entire profile represents two sequences of development; the lower involves the removal of carbonates and the translocation and accumulation of clay, while the upper involves mainly the translocation and accumulation of organic matter and sesquioxides.

## Genesis of the Eluviated and Illuviated Horizons

Although considerable information has been obtained on the morphological and chemical characteristics of the major kinds of Canadian Podzols, and about the environment in which they occur, much less is known about the processes and the mechanisms involved in their formation. The close connection between the movement of the decomposition products of organic matter and the sesquioxides has long been recognized. It is also known that, before an appreciable translocation of organic compounds and sesquioxides can take place, a considerable fraction of the exchangeable cations in the upper part of the profile must be replaced by $H^+$. This unsaturation process may take place rapidly in some mineral materials but it may be very slow in others. In calcareous materials, the carbonates must be removed and this is followed by the dispersion and translocation of clay before an appreciable translocation of organic matter and sesquioxides takes place.

Some of the early workers suggested that the organic acids brought about the solution of sesquioxides, which then moved to lower levels in ionic solutions where a higher reaction caused their precipitation. It is now recognized that the solubility of ferric iron above pH 3.5 is negligible and that most Podzols do not reach this low reaction. Under reducing conditions, however, ferrous iron may remain in solution at pH ranges found in most Podzol soils. It is therefore possible that in the Gleyed Podzols, which are saturated with water for considerable periods, some of the iron moves down the profile in ferrous form and is oxidized during periodic desiccation. In the well-drained soils, reducing conditions generally do not exist and it is therefore unlikely that this mechanism of translocation plays an important part.

Aarnio (1) was the first to suggest that hydrated free oxides are transported as negatively charged humus-protected sols. This theory assumes that the oxides are precipitated in the B horizon by divalent cations. The supply of divalent cations in the B horizons of most Podzols is, however, very limited and it is highly questionable that they could bring about the precipitation. Deb (11) has demonstrated that humus-protected sols could not be precipitated in a soil treated with calcium, and consequently the validity of this theory may be questioned.

Recently a number of investigators (2, 4, 9, 10, 14) have produced evidence to show that the translocation of organic matter and iron is brought about by the formation of mobile organo-mineral complexes. Bloomfield's investigations (5–7) with leaf extracts indicated that the organic compounds brought about the reduction of ferric iron and the formation of stable organic complexes with ferrous iron. Schnitzer and DeLong (15), however, were not able to substantiate Bloomfield's conclusions. It has been suggested (3, 13, 16) that the decomposing organic matter produces chelating substances that combine with the iron and aluminum in the soil. Though proof of the presence of chelated compounds in the soil is difficult to establish, Wright and Levick (17) have produced artificial Podzols in the laboratory with known chelating agents such as EDTA.

A more precise knowledge of the nature of the reacting organic compounds is essential before the mechanism of the interaction between organic matter and metals can be satisfactorily explained. Yarkov (19) suggested that the formation of fulvic and other simple organic acids is essential for complex formation. Gallagher and Walsh (12) suggested that simple organic acids, such as oxalic, are involved, while Bloomfield (8) attributed complex formation to the joint action of carboxylic acids and polyphenols. In a study of the nature of the organic fractions involved in the movement of sesquioxides, Wright and Schnitzer (18) have recently found that sixty per cent of this organic matter was composed of carboxyl, hydroxyl and carbonyl groups, which are known to react with metals such as iron and aluminum. It is probable that all of these reacting groups play some part in the formation of the mobile organo-mineral complexes.

Although complex formation is probably involved in the development of most Podzol soils, there is good morphological and some chemical evidence to show that, in the formation of Humic Podzols, a large percentage of the organic matter moves to the B horizon simply by infiltration. This process probably also applies, to a more limited extent, to some of the other Podzol types.

It may be stated in conclusion that during the last twenty years considerable information has been gained about the morphological and chemical characteristics of Canadian Podzol soils. More precise information is still required, however, particularly on the conditions that influence or control the development of specific soil characteristics and on the reactions that are involved. Such information would make it possible to interpret soil characteristics more reliably and would permit more accurate predictions as to how a particular soil might react to specific ameliorating treatments.

## REFERENCES

1. AARNIO, B. (1913). Experimentelle Untersuchungen zur Frage der Ausfallung des Eisens in Podsolboden. Intern. Mitt. Bodenk. 3: 131–40.
2. ALEKSANDROVA, L. N. (1954). The nature and properties of the products of reaction of humic acid and humates with sesquioxides. Pochvovedenie no. 1, 14–29.
3. BARSHAD, I. (1955). Soil Development: in F. E. Bear, ed., Chemistry of the soil. New York: Reinhold Publ. Co., pp. 1–52.

4. BECKWITH, R. S. (1955). Metal complexes in soils. Australian J. Agr. Research 6: 685–98.
5. BLOOMFIELD, C. A. (1954). A study of podzolization: III. The mobilization of iron and aluminium by Rimu (Dacrydium cupressinum). J. Soil Sci. 5: 39–45.
6. —— (1954). A study of podzolization: IV. The mobilization of iron and aluminium by picked and fallen larch needles. J. Soil Sci. 5: 46–9.
7. —— (1954). A study of podzolization: V. The mobilization of iron and aluminium by aspen and ash leaves. J. Soil Sci. 5: 50–6.
8. —— (1957). The possible significance of polyphenols in soil formation. J. Sci. Food Agr. 8: 389–92.
9. BREMNER, J. M.; MANN, P. J. G.; HEINTZE, S. G.; and LEES, H. (1946). Metallo-organic complexes in the soil. Nature 158: 790–1.
10. BROADBENT, F. E., and OTT, J. B. (1957). Soil organic matter-metal complexes: I. Factors affecting retention of various cations. Soil Sci. 83: 419–27.
11. DEB, B. C. (1949). The movement and precipitation of iron oxides in podzol soils. J. Soil Sci. 1: 112–22.
12. GALLAGHER, P. D., and WALSH, T. (1943). The solubility of soil constituents in oxalic acid as an index of the effects of weathering. Proc. Roy. Irish Acad. 49B: 1–26.
13. HIMES, F. L., and BARBER, S. A. (1957). Chelating ability of soil organic matter. Soil Sci. Soc. Amer. Proc. 21: 368–73.
14. KONONOVA, M. M. (1956). Humus der Hauptbodentypen der UdSSR, seine Natur und Bildungsweisen. Acad. Sci. URSS., Rappt. 6e Congr. Intern. Sci. Sol., 2e Comm. (Publ. in Moscow by Academie des Sciences de L'URSS).
15. SCHNITZER, M., and DELONG, W. A. (1955). Investigations on the mobilization and transport of iron in forested soils: II. The nature of the reaction of leaf extracts and leachates with iron. Soil Sci. Soc. Amer. Proc. 19: 363–8.
16. SWINDALE, L. D., and JACKSON, M. L. (1956). Genetic processes in some residual podzolised soils of New Zealand. Trans. Intern. Congr. Soil Sci. 6th Congr. Paris. E, 233–9.
17. WRIGHT, J. R., and LEVICK, R. (1956). Development of a profile in a soil column leached with a chelating agent. Trans. Intern. Congr. Soil Sci. 6th Congr. Paris. E, 257–62.
18. —— and SCHNITZER, M. (1959). Oxygen-containing functional groups in the organic matter of a podzol soil. Nature 184: 1462–3.
19. YARKOV, S. P. (1956). Seasonal dynamics of certain soil-forming processes. Pochvo-vedenie no. 6, 30–44.

# GENESIS AND CHARACTERISTICS OF
# SOLONETZIC SOILS

With particular reference to those in Alberta, Canada

## W. Earl Bowser

THE WORD "SOLONETZ" is of Russian origin and was used with reference to soils that were saline and/or alkaline. In Canada the soils of the Solonetzic Order are at present grouped as Solonetz, Solodized Solonetz, and Solod—divisions that are considered to represent stages in development or maturity. All of these have a distinctive morphology which developed, primarily, as a result of a particular characteristic of the parent material from which they were formed. In discussing these soils, however, one other Group must be considered, namely, the Solonchak. These soils are saline throughout the profile and show relatively no horizon differentiation. They have changed little morphologically from when originally laid down by geological agents. In fact, the Groups of the Solonetzic Order are believed to have developed from Solonchak, the initial salinity being the motivating agent for the genetic processes responsible for the development of these soils and for their subsequent morphology. There are extensive areas of Solonetzic soils in Western Canada.

The most common salts in the soils of Western Canada are the sulphates of sodium, magnesium, potassium, and calcium; carbonates, bicarbonates, and chlorides may also be present in small amounts. In Canada a soil is considered to be mildly saline if it contains approximately from 0.2 to 0.7 per cent of water-soluble salt, and strongly saline if over that amount. These figures would be raised if much calcium sulphate were present, and lowered if significant amounts of sodium carbonate were present. Calcium and magnesium carbonates occur in most of the soils of Western Canada; due, however, to the low solubility of these carbonates they are not considered as saline. The above noted salts were, presumably, present in the original Solonchaks.

### Occurrence

Solonetzic soils occur in many parts of the world. Some of the more extensive areas are in the southern portion of the U.S.S.R., the eastern Balkans, Australia, the less humid parts of South America, the southwestern and north-central part of the United States, and in Western Canada. There are approximately 10 to 15 million acres of Solodized Solonetz soils in Western Canada, together with a substantial acreage of Solonchak, Solonetz,

165

and Solod soils. The three largest areas of Solonetzic soils in Canada are:
the level plain between Avonlea, Weyburn, and Estevan in southeastern
Saskatchewan; a shallow till to modified bedrock area immediately north
of the International Boundary on the Saskatchewan-Alberta border; and a
relatively level area of shallow ground moraine extending some three
hundred miles from Vegreville to Taber in Alberta. Smaller areas occur in
the lacustral clays of the Red River Valley and the till plain north of Virden,
both in Manitoba; the Kerrobert area in Saskatchewan; and in the Wetaski-
win-Edmonton-Westlock area of Alberta (see Figure 1).

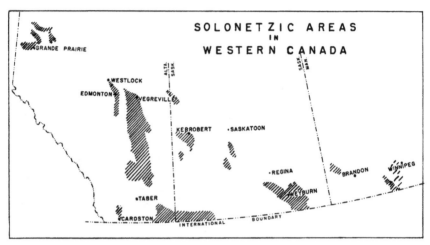

FIGURE 1.   Solonetzic areas in Western Canada

Almost all of the Solonetzic soils in Western Canada are in the semi-arid
to subhumid grassland and parkland regions. They most commonly occur
on relatively level topography in which the glacial till or other recent
deposits over the underlying bedrock is of very shallow depth. Those
occurring in fine-textured glacial lacustrine material are often near the
edges of the original lake basin. All occur in areas where internal drainage
has been or is restricted and which in most cases could have received
run-off water from surrounding higher land. In general, the areas are in
juxtaposition to Cretaceous shales which contributed to the parent material
of these soils; these shales are saline. The combination of the factors listed
above could have produced, in the immediate post-glacial period, large
salinized areas that had relatively impermeable strata fairly close to the
surface, that is, large areas of Solonchak-like soils.

### Morphology

Soils with a high salt content are usually in a flocculated (open granular)
condition and have very little vegetative cover. Under arid conditions, or

in the presence of a high water-table, the salts remain in the profile. If, however, the water table is lowered, if drainage is improved and/or the leaching effect of rainfall becomes operative, the salts will tend to be progressively leached to lower levels in the profile; the soil is being desalinized. During this desalinization, particularly if appreciable amounts of sodium are present, the soil becomes strongly alkaline. This process, known as solonization, causes certain morphological changes to take place in the profile. The first product of this change is the development of a Solonetz.

The Solonetz is characterized by a relatively thin surface (A) horizon that varies in structure from hard lumpy to fairly friable and granular. The latter type of structure is found in the better drained positions on which grass has become established. Immediately below the A, there is a relatively thick columnar (B) horizon. The columns vary from one to four inches in diameter and are often pointed at the top. The B horizon contains more clay than the A horizon; it is very hard when dry and very plastic when wet. The surface of the columns is usually darker in colour than the interior, primarily due to a thin coating of reprecipitated organic matter. Few roots penetrate into the columns; what roots do go down follow the cleavage lines between the columns. Below the B horizon, there is usually a measurable accumulation of salts and lime carbonate. In the semi-arid Brown zone of Alberta, this accumulation starts at about twelve to eighteen inches from the surface; in the subhumid Black zone, it starts at twenty to thirty-six inches. This salt accumulation horizon gradually gives way to the undifferentiated parent material below (the C horizon).

With continued leaching, a process of dealkalization or solodization takes place and the Solonetz becomes a Solodized Solonetz. These soils have a dark upper A horizon that usually contains approximately the same amount of organic matter (humus) as the adjacent Chernozemic soils.

FIGURE 2. A Dark Brown Solodized Solonetz. Note the difference in structure between horizons.

The lower part of the A horizon, however, is much lighter-coloured than the upper part and is often of a platy structure. It is leached and low in clay content. The hard columnar B horizon has white-capped rounded tops and the faces of the columns are not as dark-coloured as in the Solonetz due in part to a coating of silica grains washed down from the leached horizon above. The upper part of the B horizon is quite firm, but can usually be crushed by hand into angular blocks; the very hard B horizon of the Solonetz is beginning to break down so that roots more readily penetrate into the columns. Below the columnar B, there is usually a compact but less structured horizon transitional to the lime and salt accumulation horizon. As in the Solonetz, there is a gradual change to the C horizon below.

With increased solodization the Solodized Solonetz gives way to a Solod. In so doing the A horizon becomes relatively deeper, the B horizon relatively thinner, and the salt moves to lower levels. The remnant outlines of the former columnar B can often be seen in the present A horizon. The remaining B horizon crushes readily into a small to medium blocky structure. In contrast to the Solodized Solonetz, there is usually a pronounced AB transitional horizon. This horizon has the pale brown colour of the leached portion of the A, and the blocky structure of the B. The inside portions of these blocky aggregates are usually darker than the surfaces. A typical Solod from the Black Soil zone of Alberta has from six to ten inches of black friable surface A; from three to six inches of very pale brown friable and platy subsurface A; from one to two inches of pale brown blocky AB; from six to twelve inches of small to medium blocky B that is well penetrated with roots; and from six to twelve inches of lower B that is medium to large blocky and less friable than the upper B. Below this is a lime carbonate accumulation horizon that also contains some salt although usually in much smaller amounts than in the Solodized Solonetz soils.

In Western Canada, Solonetzic soils in different stages of development occur in close association. The usual sequence, if there is a topographic and therefore a drainage variability, is for the Solonchak to occur in the lowest position, followed, up slope, by the Solonetz, then the Solodized Solonetz and Solod. If considerably higher ground occurs, that is a surface much farther removed from an impermeable subsoil, it usually has Cherno-zemic (non-saline) soils. In level areas one Group dominates, unless there is a major change in parent material.

A distinctive micro-relief is often found in Solodized Solonetz areas. A differential erosion has taken place so that a percentage of the area has lost the A horizon, producing eroded pits with steep, sharply defined edges and usually sparsely covered with vegetation. These may make up from 10 to 50 per cent of the area. The salt content of the B horizon in these locations is usually higher than in the B horizon of nearby Solodized Solo-netz soils where the A horizon is still intact. The origin of these eroded pits is still a matter of speculation. Observation would suggest that there was

FIGURE 3. Uneven growth of wheat in a Solonetzic area. The poorest growth is where the A horizon is very shallow. Note the sharp line between the A and B horizons in the road cut.

originally a morphological difference which could have initiated a mechanical removal.

## Genesis and Properties

Considerable research has been conducted on Solonetzic soils both in the field and in the laboratory; research designed to determine the processes (the genesis) by which these soils have developed. There are many published papers on the subject; it is not proposed, however, to give here an extensive bibliography. Bentley and Rost (1) give a fairly complete one up to 1945; Ehrlich and Smith (3) report work up to 1957. In 1956, the Commonwealth Bureau of Soils (2) published a bibliography, with short abstracts, reporting work done primarily in Russia and the Balkan states between 1927 and 1956. The writer has drawn on information in the reports of the Manitoba, Saskatchewan, and Alberta Soil Surveys, which are herewith acknowledged.

As already suggested, all the soils of the Solonetzic Order are believed to have developed from saline parent material, and the morphological changes that have taken place during their progressive development were triggered by the process of desalinization. The initial stage, the Solonchak, has an undifferentiated profile with salt distributed throughout its depth. There is no development of pedogenic horizons; due to the high salt content, the soil is usually in a flocculated condition. As the salts are leached out, their buffering effect is removed and if, at this time, over 10 to 20 per cent of the exchangeable cations on the complex is sodium the soil becomes deflocculated (highly dispersed). At this point it is believed that there is hydrolysis of the sodium clay producing hydrogen clay and sodium hydroxide or sodium carbonate; both of the latter are soluble and highly alkaline. This process was suggested by K. K. Gedroiz in 1912 (4). Under conditions of high alkalinity aluminum, iron, silica, and humic acids go into solution. The percolating rain water would then carry the products in solution, as well as finely divided material in suspension, down to lower depths. This material, it is suggested, would move down until it reached a point where the salt content was sufficiently high to cause reflocculation or until the pH was such that secondary clays would be resynthesized. This transported and re-deposited material would then tend to close the pore spaces and form a semi-compact layer of low permeability. As more material was carried down, it would then tend to accumulate above the previously formed layer.

Certain workers have suggested that a high magnesium content could have produced an effect similar to that of sodium and the term magnesium solonetz has been used. It is known that the magnesium ion has some deflocculating effect on soil. The products of hydrolysis in this case would, however, be much less soluble than is the case with sodium and the resultant media would be less alkaline. It is of interest to note that in many areas, where magnesium has been suggested as the triggering cation, there is a high to fairly high content of the chloride anion. In Alberta, where sodium appears to be the responsible cation, the principal anion is sulphate and there are practically no chlorides. It has been suggested by Joffe (6) that magnesium has an additive effect to sodium.

The process outlined above is generally accepted as the process of solonization and the product as a Solonetz. The hard columnar B is the significant morphological characteristic of this soil. These columns are possibly caused by alternate wetting and drying common to the dryer climates. Most of the clay in the soils of Western Canada is of the montmorillonitic and illitic type and some of these soils have a swelling index of up to 50 per cent. Mechanical analyses show that in Alberta there is from 10 to 50 per cent more clay in the B horizon of the Solonetz than in either the overlying A or underlying C horizons.

With the removal of salt from the surface, the vegetative cover increases. The humus content of the surface or A horizon increases, it usually becomes more friable, and the pH may drop considerably. Analyses of the A horizons of eight Alberta Solonetz soils gave a pH range of 5.3 to 6.2. This low pH may be due in part to aluminum. The parent material, C horizon, is saline and in Western Canada also contains, on the average, 2 to 5 per cent calcium-magnesium carbonate.

It is thought that the second stage through which the soils of the Solonetzic Order may progress, namely, dealkalinization or solodization, begins at an early stage. The pH values reported above would suggest this since, in the initial hydrolysis of the sodium clay, a much higher pH would be expected. Although these soils have the morphology associated with a Solonetz, chemically they are already approaching a Solodized Solonetz. With an increase in vegetative growth, there would be an increase in the amount of humic acids formed and also in the return of bases, principally calcium, to the surface. A low pH would also tend to render the alkali earth carbonates more soluble. Calcium and magnesium could then dominate the exchange complex.

Percolation studies made on Solodized Solonetz soils in the Brown zone of Alberta indicate that water infiltration into the B is very slow. When three inches of water were allowed to stand on the top of the B horizon for twenty-four hours, it only penetrated from one-half to two inches into this horizon. Percolating rain water, containing humic acids and silica from plant decomposition, would therefore be temporarily arrested at the top of the B horizon. This AB contact, then, could well be the locale for the most intense weathering. The solodization process is characterized by the formation at the AB contact of a very light-coloured eluviated horizon relatively low in clay. Concurrently, there appears to be a breakdown of the columnar B from the top downwards and the columns develop rounded white tops. As the degradation and eluviation continue, the depth of the leached horizon increases and the length of the columnar B horizon decreases. Some degradation also appears to take place throughout the columns and roots penetrate readily into them.

Mechanical analyses of 14 Alberta profiles showed that six upper B horizons had less total clay than the C (from 10 to 30 per cent less) and eight had more clay (from 10 to 50 per cent). Five of the lower B horizons had less clay (from 10 to 30 per cent) and nine had more clay (from 10 to 100 per cent) than the underlying parent material. It is possible that in those cases where there was a lower clay content in the B horizon than in the C horizon, there were originally variations in the parent material or the eluviated products were distributed over a fairly great depth. The "average" depth of rain penetration in the semi-arid to subhumid regions of Western Canada is considered to be from two to four feet. This is the average depth to the lime and salt accumulation horizon. Deep drill sampling, however, done at Vauxhall, Alberta, by the Drainage Division of the Prairie Farm Rehabilitation Administration of the Department of Agriculture (Canada), has shown that in the semi-arid Brown zone some salt has been moved down nearly ten feet. It appears reasonable to assume that the products of weathering during the solodization process could also be distributed over much of this depth. In all profiles there was from 100 to 300 per cent more clay in the B horizon than in the leached portion of the A horizon.

The leached, or illuviated, portion of the A horizon develops a characteristic platy structure during the solodization process. This platy structure may be due, in part, to the lateral growth of roots above the hard B horizon or to the water temporarily held in this zone, particularly during periods when the temperature fluctuates around the freezing point. These plates are often darker-coloured on the under side than on the top. The pH of the A horizon of the Solodized Solonetz is neutral to slightly acidic. The pH of the top of the B horizon, that is, the portion that is most strongly weathering, is often the most acidic and may be below 5.5. The main B horizon is slightly acidic to somewhat alkaline, usually from pH 6.0 to 8.0. The exchange complex of the B horizon is primarily base-saturated. In Alberta this horizon has usually from 10 to 30 per cent exchangeable sodium; in Saskatchewan Jansen and Moss (5) report from 10 to 40 per cent sodium; in Manitoba Ehrlich and Smith (3) report under 5 per cent sodium but from 40 to 50 per cent exchangeable magnesium. The vegetative cover on these soils is usually fairly typical of the region. At this point the soil is starting to regrade, or to take on more of the characteristics of the non-saline Chernozemic soils of the region. The intermediate stage between the Solodized Solonetz and the regraded Chernozem is the Solod.

In this regrading process, the dark-coloured surface horizon (containing humus) appears to follow the eluviation down. Bleached silica grains, possibly from the former leached portion of the A horizon, can be seen mixed with the present darkened surface. The B horizon loses its pronounced columnar structure and the ratio of clay in the B to the C approaches one. Analyses of six Solod soils showed one with 10 per cent less clay in the B than in the C, one with no difference, and the other four with between 10 to 20 per cent more. As in the Solodized Solonetz soils, there is more clay in the B than in the leached A. Out of twelve Alberta profiles analysed, one showed no difference and the other eleven had from 50 to 100 per cent more clay in the B than in the A. The eluviated lower A has, in most cases, a lower clay content than the upper portion of the A.

The pH of Solod* soils in Alberta range from 6.0 to 7.0 in the upper A, 6.0 to 6.5 in the lower A, 5.5 to 7.0 in the B, and 7.0 to 8.0 in C horizon. Sodium usually makes up less than 10 per cent of the exchangeable bases in the A and B horizons, and magnesium from 20 to 40 per cent. The remainder is primarily calcium.

The soils of the Solonetzic Order, in Canada, as described above, have been those found in the grassland region. Soils with related morphology have been described by Odynsky (7) in portions of the forested region of the Peace River area of Alberta. The parent material of most of these soils is a shaley till derived from the underlying Alberta shales of Cretaceous age. These are saline. Some of these soils in the lower horizons still have a high percentage (10 to 30 per cent) of exchangeable sodium. They are now under tree cover and the surface horizons have taken on the characteristics of the surrounding Grey Wooded soils (a Group of the Podzolic Order). It is suggested that the tree invasion was possibly delayed until the salts were sufficiently leached out and the pH sufficiently reduced to produce a habitat

*Some of these are now classed as Eluviated Chernozemic soils.

that could be tolerated by the aspen poplar and birch which make up most of the forest cover. The podzolic degradation could then have been superimposed on the solodized profile.

It is not known whether Solonetzic soils are still developing or whether most have reached a static stage. It is quite possible that, under the limited rainfall which now prevails over much of the area, an equilibrium has been reached and that quite a drastic change of environment might be necessary to carry the process forward.

## Economic Significance

At least three characteristics of Solonetzic soils are of significance in their management. These are the hard, relatively impermeable, B horizon; the salt content; and the restricted subprofile drainage.

The Solonetzic B horizon limits water and root penetration. Rain water (or added irrigation water) remains on the surface rather than infiltrating into the soil and hence much of it is lost by evaporation. Roots tend, therefore, to concentrate in the A horizon. As a result, crops on Solonetzic soils suffer more during extended dry periods than do those on the more permeable Chernozemic soils. Salts have a depressing effect on plant growth. This effect is particularly noticeable in the Solonchak and Solonetz Groups. The physiologic effect of the interrelation of the various soluble ions may also be highly significant. Considerable research is now being done in relation to the agricultural use of these soils. Such things as improved methods of cultivation, the addition of chemical amendments, and the selection of more adaptable crops are all receiving attention.

These soils also present problems in the field of engineering. The corrosive effect of salt on materials used in construction is well known. Soils that have a high content of sodium on the exchange complex tend to be deflocculated and erode readily. This affects their usability for earthen structures. Since most of these soils are underlain by strata of low permeability, drainage problems are often encountered.

### REFERENCES

1. BENTLEY, C. F., and ROST, C. O. (1947). A study of some solonetzic soil complexes in Saskatchewan, Canada. Sci. Agri. 27: 293.
2. Commonwealth Bureau of Soils, Farnham Royal, Bucks, England (Sept. 1956). European work on solonetz soils and their improvement (1927–1956). Mimeographed release.
3. EHRLICH, W. A., and SMITH, R. E. (1958). Halomorphism of some clay soils in Manitoba; Can. J. Soil Sci. 38: 103.
4. GEDROIZ, K. K. J. Expt. Agron. (Russian, 1912), 13: 363; quoted in Russel, Soil Conditions and Plant Growth (Longmans, Green and Co.).
5. JANSEN, W. K., and MOSS, H. C. (1956). Exchangeable cations in solodized solonetz and solonetz-like soils in Saskatchewan; J. Soil Sci. 7: 203.
6. JOFFE, J. S. (1949). Pedology. New Brunswick, N.J.: Pedology Publications.
7. ODYNSKY, W., WYNNYK, A., and NEWTON, J. D. (1952). Soil survey of the High Prairie and McLennan sheets. Report 17, Alberta Soil Survey, University of Alberta.

# THE SOILS OF SOUTHERN ONTARIO

## N. R. Richards

SOUTHERN ONTARIO includes that portion of the province lying beween 75 degrees and 83 degrees west longitude and 42 degrees and 46 degrees north latitude, and so the region lying south of the French and Mattawa Rivers and Lake Nipissing. The total land area is approximately 51,000 square miles (32,656,000 acres). Slightly more than 13 million acres are part of the Canadian Shield. Approximately 55 per cent is occupied farm land, most of which occurs south of the Shield. For more than forty years, soil surveys have been conducted in Ontario. The surveys for the most part have been co-operative projects between personnel of the Ontario and Canada Departments of Agriculture. The headquarters for the personnel associated with soil survey projects has been the Department of Soil Science, Ontario Agricultural College, Guelph. Information on the characteristics of the soils of Southern Ontario is described in published and unpublished reports of the Ontario Soil Survey.

### NATURAL DIFFERENCES AMONG SOILS

More than five hundred soil types have been classified and mapped in Southern Ontario, each of which has a combination of characteristics peculiar to itself and which distinguishes it from other soils. The existence of differences and similarities, expressed to various degrees in the soil profile, provides the basis for the classification of soils. The characteristics of the different soils reflect the effect produced by several soil-forming factors namely climate, vegetation, the nature of the mineral parent material, relief and drainage, and the length of time these factors have been active (age).

The fundamental unit in the system of soil classification is the soil type. On the basis of selected characteristics in the profile the soil types are classified into the higher categories of Subgroup, Great Soil Group and Order. The dominant characteristics of the following Orders, Great Soil Groups, and Subgroups are discussed with particular reference to Southern Ontario.

| ORDER | GREAT SOIL GROUP | SUBGROUP |
|---|---|---|
| Brunisolic | 1. Brown Forest | 1a. Orthic Brown Forest |
| | | 1b. Degraded Brown Forest |
| | | 1c. Gleyed Brown Forest |
| | 2. Acid Brown Wooded | 2a. Orthic Acid Brown Wooded |

174

Podzolic                3. Grey Brown Podzolic  3a. Orthic Grey Brown
                                                      Podzolic
                                                 3b. Brunisolic Grey
                                                      Brown Podzolic
                                                 3c. Gleyed Grey Brown
                                                      Podzolic
                        4. Grey Wooded           4a. Gleyed Grey Wooded
                        5. Podzol                5a. Orthic Podzol
                                                 5b. Bisequa Podzol
                                                 5c. Gleyed Podzol
Gleysolic               6. Dark Grey Gleysolic   6a. Orthic Dark Grey
                                                      Gleysolic
Regosolic               7. Regosol
Organic

The distribution of the Great Soil Groups and Subgroups is shown in Figure 1. Each area naturally contains an association of several Great Soil Groups and Subgroups but within each area, one Great Group and one Subgroup is dominant.

## DOMINANT CHARACTERISTICS OF THE ORDERS, GREAT SOIL GROUPS, AND SUBGROUPS

### Brunisolic Soils

The Brunisolic soils developed under forest cover. In these soils the process of soil leaching has not resulted in the formation of distinct zones of translocation and accumulation. The profiles, beneath the surface horizon are dominantly brown in colour. Two Great Soil Groups occur in this Order in Southern Ontario, the Brown Forest and the Acid Brown Wooded.

1. *Brown Forest Soils*  The Brown Forest soils have developed under deciduous trees on calcareous materials of good to imperfect drainage. The A horizon is a dark friable and granular mineral organic surface which grades into a friable brown B horizon underlain by the C horizon which is calcareous and occurs at depths of fifteen to eighteen inches. The surface soil is usually about neutral to slightly alkaline in reaction and saturated with exchangeable bases. These are immature soils and, given time under similar vegetation, can be expected to develop into Grey Brown Podzolic soils. The various subgroups that have been recognized are as follows:

1.(a) *Orthic Brown Forest Soils*  These are well drained soils; the $A_1$ horizon is four to five inches thick; there are no marked eluvial or illuvial horizons; the B horizon ranges in depth from six to ten inches.

1.(b) *Degraded Brown Forest Soils*  The Degraded Brown Forest soils differ from the Orthic due to the presence of a weakly developed illuvial horizon; there is some clay accumulation in the B horizon.

1.(c) *Gleyed Brown Forest Soils*  These are imperfectly drained soils and the mottled appearance due to the presence of hydrated iron compounds is present throughout the profile; the $A_1$ horizon is thicker than in the Orthic group.

2. *Acid Brown Wooded Soils (Brown Podzolic)* The Acid Brown Wooded soils have developed in materials with low base saturation (less than 50 per cent). Below the $A_0$ horizon, which is a thin leaf-mat, is a B horizon ten to twenty inches thick, brown or reddish-brown, and moderately to strongly acid.

2.(a) *Orthic Acid Brown Wooded Soils* This is the only subgroup of Acid Brown Wooded Soils that has been classified in Southern Ontario: these are well-drained soils with low base-saturation, the profile characteristics being essentially the same as those described for the Acid Brown Wooded Great Soil Group.

*Podzolic Soils*

Most of the soils in Southern Ontario have developed under a forest vegetation. The downward movement of the soil solution containing the products of decomposition of the leaf litter has leached the soil of many of .its original compounds. These soils are classified as Podzolic soils and are characterized by a grey layer occurring immediately beneath the surface organic layer which in turn is underlain by a darker subsurface horizon enriched in clay or organic matter and sesquioxides. The Grey Brown Podzolic, Grey Wooded and Podzol Great Soil Groups are Podzolic soils.

3. *Grey Brown Podzolic Soils* Soils of this Great Soil Group have developed in materials with high base content. The $A_1$ horizon is a friable, mineral organic layer, the $A_2$ horizon is lighter in colour and is underlain by the B horizon which has a darker colour and is a prominent textural layer. The following subgroups occur:

3.(a) *Orthic Grey Brown Podzolic Soils* These soils are well drained, the profile being characterized by a surface horizon four to five inches thick, underlain by a grey $A_2$ horizon four to twenty inches thick; the B horizon is well developed and usually ranges in thickness from four to ten inches and contains more sesquioxides and clay than the other horizons in the profile.

3.(b) *Brunisolic Grey Brown Podzolic Soils* The main difference between this group and the Orthic group is that the upper $A_2$ horizon is brown rather than grey; the profile in the upper A horizon exhibits the characteristics of the Brunisolic soils.

3.(c) *Gleyed Grey Brown Podzolic Soils* The Gleyed Grey Brown Podzolic soils have developed in materials of high base status but are less well drained than the Orthic and Brunisolic groups; the $A_1$ horizon, four to six inches thick, is a dark friable mineral-organic layer, slightly acid to neutral, overlying the eluviated layer which is mottled; the B horizon is similar in textural characteristics to the illuviated horizon of the Orthic group; the colour of the horizons below the $A_1$ is duller than that found in the Orthic and Brunisolic subgroups; the gleyed soils occur in association with the Orthic and Brunisolic subgroups.

4. *Grey Wooded Soils* The Grey Wooded soils have developed under mixed hardwood and conifer vegetation from calcareous parent material. They have a relatively thick organic surface layer over a distinct grey leached layer which in turn is underlain by a darker subsurface horizon enriched in clay. Orthic, Bisequa and Gleyed Grey Wooded soils have been mapped in Ontario. Only the Gleyed Grey Wooded subgroup occurs in an area extensive enough to be shown in Figure 1.

4.(a) *Gleyed Grey Wooded Soils* The $A_0$ horizon is a thin leaf-mat one to two inches thick; the grey $A_2$ horizon is mottled, and is underlain by a textural B horizon; the B horizon is a marked illuviated layer: the surface and subsurface layers are slightly to moderately acid.

The Orthic Grey Wooded soils are well drained; in the Bisequa Grey Wooded soils a profile with the characteristics of an Orthic Podzol occurs in the A horizon.

5. *Podzol Soils* The distinguishing characteristics of the Podzol Great Soil Group is a well-defined, strongly weathered and leached grey layer immediately below the surface horizon. The accumulation products in the subsurface consist dominantly of organic matter and sesquioxides. The Podzol soils are moderately to strongly acid and consequently have a low base saturation.

5.(a) *Orthic Podzols* These are well-drained soils; the $A_0$ horizon is approximately one inch thick; the grey $A_2$ horizon is two to four inches thick and overlies the brown B horizon which ranges in depth from six to twenty inches; the main accumulation products in the B horizon are organic matter and/or iron.

5.(b) *Bisequa Podzols* In the Bisequa Podzols a profile similar to the Orthic Podzols occurs in the upper eighteen inches of the profile; the orthic-like profile overlies the B horizon of a Grey Brown Podzolic or Grey Wooded soil which is found at depths of thirty inches or more; the structure of the underlying B horizon may be slightly altered depending upon the stage of break down or decay it has reached.

5.(c) *Gleyed Podzols* These are imperfectly drained soils; the $A_0$ horizon is one to two inches thick; the grey $A_2$ horizon two to six inches thick is mottled and overlies a mottled dark reddish-brown B horizon; the distinguishing characteristic of the group is the well-defined eluviated mottled layer.

*Gleysolic Soils*

The Gleysolic soils are poorly drained. The normal processes of development have been restricted due to high water-tables. Beneath the surface layer the subsurface is dull-coloured and mottled, a result of the dominantly reducing conditions that have prevailed in the soil materials.

**6.(a)** *Orthic Dark Grey Gleysolic Soils*   A dark grey $A_1$ horizon five to six inches thick is underlain by gleyed layers without apparent eluviation or illuviation; the gleyed layers are mottled with brownish rusty or bluish spots, streaks and splotches.

### Regosolic Soils

These are immature soils without any distinct colour or horizon differentiation in the profile. Profile development is mainly confined to the surface layer.

**7.** *Regosol Great Soil Group*   This is the only Great Soil Group that has been mapped in the Regosolic order in Southern Ontario; the soils lack horizon development due to the nature of the parent material; the grey brown $A_1$ horizon two to three inches thick is underlain by grey fine textured materials which are practically unaltered.

### Organic Soils

The decayed or partially decayed plant materials from which the organic soils of Southern Ontario were formed are considerably more decomposed than the peat bogs of Northern Ontario. These soils do not possess the characteristic horizon differentiation which is typical of the profile development in the mineral soils. Most of the organic soils in Southern Ontario have developed from swamp grasses and sedges and are very dark grey muck. In the Canadian Shield area the raw nature of the organic deposits is due to the slow rate of decomposition of the moss in the vegetation from which they developed.

### SOIL AND AGRONOMIC USE PATTERN IN SOUTHERN ONTARIO

In Figure 1 the distribution of dominant Great Soil Groups and Subgroups is shown for Southern Ontario. Each area consists of an association of several groups. The extent and pattern of distribution of the various groups determine to a large extent the use that is made of the soils within an area. The characteristics of the various Great Soil Groups and Subgroups have been discussed previously. The pattern that the Subgroups assume in Southern Ontario and some general observations concerning land use in the several areas in Figure 1 follow.

*Area 1a*   Orthic Brown Forest soils are dominant in this area and occur in association with Degraded Brown Forest, Gleyed Brown Forest, Orthic Dark Grey Gleysolic and Organic soils in the northern part of the Bruce Peninsula, Manitoulin Island, Prince Edward County and in several areas in eastern Ontario. The soil deposits are usually less than three feet over limestone bedrock. Much of the land in this area is used for grazing purposes. Some areas are very stony and the limestone bedrock is often exposed.

*Area 1b*   Degraded Brown Forest soils occur in association with Orthic and Gleyed Brown Forest soils along with lesser proportions of Orthic Dark

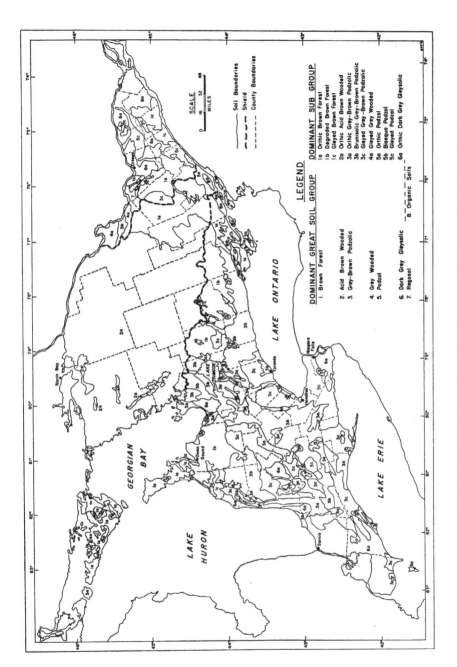

FIGURE 1.  Distribution of dominant Great Soil Groups and Subgroups in Southern Ontario

Grey Gleysolic and Organic soils. These soils are found south of the Canadian Shield in central Ontario and to a lesser extent in the Bruce Peninsula. Livestock farming is the main enterprise found in this area. The materials from which the soils developed are dominantly of glacial till origin and many of the areas are stony. The topography is often hilly and sheet erosion can be serious when the soils are cultivated.

*Area 1c* Gleyed Brown Forest soils dominate in this area along with lesser proportions of Orthic and Degraded Brown Forest, Dark Grey Gleysolic, and Organic soils. The use pattern is similar to Areas 1a and 1b, but the suitability of the soils is lessened for agricultural purposes due to the predominance of imperfectly and poorly drained soils. Drainage improvement through the use of open ditches and tile drains improves the capability of the soil for growing agricultural crops.

*Area 2a* Orthic Acid Brown Wooded soils are found in association with Orthic Podzols, Gleyed Podzols, Orthic Dark Grey Gleysolic and Organic soils. For the most part these soils are found in the area of the Canadian Shield. The soils often occur as thin veneers over the knolls and ridges of granite, quartzite and other siliceous rocks. Some agricultural enterprises are carried on in the Shield. Forestry and recreational endeavours occupy most of the area, a use for which it is well suited.

*Area 3a* The Orthic Grey Brown Podzolic soils occur most frequently in southwestern Ontario in Huron, Middlesex, Perth, Wellington and York Counties. They are the dominant soils in the area but Gleyed Grey Brown Podzolic, Orthic Dark Grey Gleysolic and Organic soils occur in association with them. Generally speaking the materials of this group are fine-textured. Farm enterprises include various combinations of production of livestock, cereal grains, hay and pasture and cash crops. On the more rolling areas, the soils are susceptible to erosion which can be controlled by keeping the area under cover for as large a proportion of time as possible or through strip cropping.

*Area 3b* The Brunisolic Grey Brown Podzolic soils occur in the heartland of southwestern Ontario and in south-central Ontario. Other soil groups in this area are the Gleyed Grey Brown Podzolic, Orthic Dark Grey Gleysolic and Organic soils. These soils are used for a wide variety of purposes, including the growing of tobacco for flue-curing, canning and vegetable crops, fruit crops, cereal grains, hay and pasture. Dairy and beef farming are agricultural enterprises commonly found on these soils.

*Area 3c* The Gleyed Grey Brown Podzolic soils are imperfectly drained and since they are found in association with the subgroups contained in 3a and 3b occur in similar geographic locations. When the drainage hazard is overcome, the Gleyed Grey Brown Podzolic soils can be used for essentially the same purposes as their well-drained associates.

*Area 4a* The Gleyed Grey Wooded soils are found in association with the Orthic and Bisequa Grey Wooded, Dark Grey Gleysolic and Organic soils. This association occurs in Manitoulin Island, Carleton and Renfrew

Counties and the southern part of Lennox, Addington and Frontenac Counties. These soils are used for livestock farming and the growing of spring grains and forage crops.

*Area 5a*    Orthic Podzols are the dominant soils found within the Canadian Shield. Other soil groups that occur in association with them are Orthic Acid Brown Wooded, Gleyed Podzols, Dark Grey Gleysolic and Organic soils. The soils in this group are not used extensively for agricultural purposes. Forestry enterprises are common. For the most part the soil materials are coarse-textured and acid in reaction.

*Area 5b*    The Bisequa Podzols occur in the Alliston area of Simcoe County. Gleyed Podzols, Brunisolic Grey Brown Podzolic, Dark Grey Gleysolic and Organic soils also occur in this area. These soils are well suited to a wide variety of uses some of which include growing tobacco, vegetable crops, cereal grains, forage crops, sod farming, and beef and dairy enterprises. The geographical location of these soils in relation to the large urban centre of Toronto makes them especially valuable for specialized types of farming.

*Area 5c*    In addition to the Gleyed Podzols, Orthic and Bisequa Podzols, Acid Brown Wooded, Dark Grey Gleysolic and Organic soils occur in area 5c. Most of these soils are imperfectly drained which limits their suitability for the crops grown on the soils found in 5b.

*Area 6a*    The Dark Grey Gleysolic soils are the dominant component of area 6a. Other groups include Organic soils, with lesser amounts of the other subgroups depending upon geographic location. These soils occur extensively in Essex, Kent, Lambton, Huron, Bruce, and Welland Counties and in eastern Ontario. Poor drainage is the main hazard associated with these soils as far as their suitability for the growing of agricultural crops is concerned. In southwestern Ontario (in Essex, Kent and Lambton Counties) tile drains and open ditches have been installed extensively. Following drainage improvment large acreages of grain corn, soybeans, canning crops and other cash crops are produced. In eastern Ontario the Dark Grey Gleysolic soils are used extensively for dairy farming enterprises. The capability of the Gleysolic soils for agricultural purposes is greatly improved when the drainage hazard is overcome.

*Area 7*    The Mull Regosols dominate this group along with Orthic and Gleyed Podzols and Muck. They occur in Carleton and Renfrew Counties. Many of the soils contained in this area are fine-textured. Spring grains and forage crops for livestock enterprises are produced extensively on the Regosols.

*Area 8*    Although there is a large acreage of organic soils in Southern Ontario, they do not occur as individual extensive deposits but are found as small areas in association with mineral soils. The organic soils have been developed for market garden purposes at Bradford, Thedford, and Alfred. The large proportion of the organic soils in Southern Ontario has not been developed for agricultural purposes.

## ACKNOWLEDGEMENTS

The basic information for this paper was obtained from information compiled by personnel associated with the Ontario Soil Survey. The author acknowledges the helpful assistance of Dr. P. C. Stobbe's paper "The Great Soil Groups of Canada" (A.I.C. Review, vol. 15, no. 2) in the preparation of this contribution.

# CORRELATION OF ENGINEERING AND PEDOLOGICAL SOIL CLASSIFICATION IN ONTARIO

## A. Rutka

THE PEDOLOGICAL SYSTEM of soil classification was developed primarily for agricultural purposes at the beginning of this century. This system was applied in the preparation of soil maps on a county-wide basis, wherein the soils were separated into units, each unit representing soils having the same texture, colour, structure depth and number of horizons. The chemical composition of the soil, along with its suitability for various crops, was also given for land use and soil management purposes.

It did not take very long for some highway engineers to realize the usefulness of the agricultural soil map to the field of highway engineering. In 1924, in his study of pavement design in the Pacific Northwest, A. C. Rose pointed out the advantages of using county soil maps, soil keys, and significant soil profile characteristics for mapping soils. Shortly thereafter, in 1925, Michigan was the first state to use the soil survey map on a state-wide basis to study and correlate pavement behaviour with soil type. In 1927 Michigan initiated the use of the soil map as a routine procedure for making soil surveys for design purposes. Slowly other states began to use the pedological method in a similar manner to Michigan and to use the agricultural soil maps to locate construction materials and to plan and organize soil survey work.

Ontario began the first extensive use of agricultural soil maps in 1945, when the Soils Section of the Department of Highways was formed. Soil maps were used in correlating pavement behaviour with soil type and with other factors that influence performance. As there had been virtually no reconstruction of highways during the Second World War, many highways had deteriorated in spite of normal maintenance and an excellent opportunity was thus available for the study of many of the factors that influence road performance. The factors considered in this study were: (1) Surface and internal drainage of the soil; (2) Height of grade with respect to ground line; (3) Depth of pavement construction, including granular base and pavement, and quality of same; (4) Type of subgrade material and its condition.

It was found that pavement performance could be directly correlated with soil type. The main problem associated with roads in the Province of Ontario is that caused by frost action. Pavements that performed poorly were noted to have subgrade and/or granular materials susceptible to frost action; these materials could be sorted out on the basis of their textural composition in most cases. Thus, as a key to the performance study and as a tool for soil surveying for highway purposes, the agricultural soil map came into use. The information procured by the pedological method of soil classification, with some additional laboratory testing of the soils and the exercise of sound judgment and experience, could be applied directly to design.

A further programme was started in 1948, entitled "The Engineering Evaluation of Engineering Soils in Southern Ontario." It was the purpose of this programme to collect several representative samples of each soil series in Southern Ontario and to determine the engineering characteristics of each of the horizons of each series. It was found, however, that there was a wide variation in test results in several of the series and the results were of limited use. Furthermore, only approximately half of the counties had been mapped at that time and the information could apply only to areas where a soil survey report had been prepared. Some attempt was made to group soil series that had similar engineering characteristics, in order to reduce the number of soil series names. This programme, although it could not be completed until the whole province was mapped, did not yield information in a form suitable for use in engineering.

The programme was revised recently, in such a way as to sample soil series on a county-wide basis. Since any particular county would be mapped by one soil scientist, there would be uniformity in soil mapping within that county. In addition to obtaining representative samples on each horizon of each soil series, further information was obtained regarding any features which could affect the construction operations. One county (Peel) has so far been sampled, but the engineering report, to be appended to the soil survey report, has not yet been completed.

Many engineers have been reluctant to accept the pedological system of soil classification, because they prefer a self-descriptive mapping unit which would give an appraisal of the soil by factors which bear directly upon the problem in hand. Efforts to formulate such a system have centred around the development of a laboratory classification system, wherein soils are differentiated upon the basis of their physical properties. With this method, it has been necessary to correlate test results with field performance; for the most part, good success has been achieved. It is evident, however, that all the classification tests are performed on the soils in a disturbed state and that environment, the most important factor, is not considered. It is well known that soils with identical constants and classifications, but located in a different environment, may react quite differently in the field when subject to equal traffic volumes and loads. Although there are many classification systems for engineering purposes now in use, a classification system for highway purposes must take into account the in-place conditions where such factors as parent material, topographic position and profile development are considered.

The object of any soil classification system for highway engineering purposes is to eliminate an extensive laboratory testing programme, and yet provide all the pertinent information that is required. The pedological classification can provide the engineer with such a system. In an area where an agricultural soil map has been prepared, the soil scientist has made a thorough study of the area and has mapped and given names to all the soil types encountered. This means that any soil type, wherever it is shown, exhibits the same profile characteristics and should, accordingly, exhibit the same physical properties within narrow limits. It is only necessary, therefore, for the engineer to obtain a few representative samples of each horizon of each soil type for laboratory testing purposes. The laboratory tests should, therefore, apply within the boundaries of the area shown on the soils map.

Some soil types occur in areas that are too small to be shown on the map and the boundaries between the various soil types are not usually given in sufficient detail for highway purposes. The highway engineer must, therefore, establish boundaries accurately, and procure additional information regarding other design requirements. In Ontario, the highway soil survey involves the preparation of a soils strip plan approximately 300 feet wide (150 feet on each side of the centre line). The soil type is determined, along with its boundary, by means of shallow hand borings (see Figure 1) and this information is shown on the strip plan. Since soil survey reports have not been prepared for many of the counties in Ontario, it is difficult

### SOILS MAP FOR GENERAL SOIL ENGINEERING INFORMATION

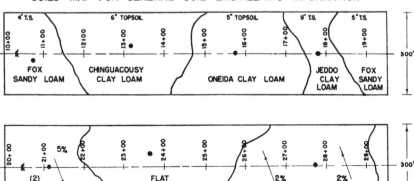

FIGURE 1.   Soil Map for General Soil Engineering Information

for the engineer to apply soil type names. The engineer can, however, distinguish one soil type from another, and can establish the boundaries, but must give a comprehensive description of each of the layers of the soil profiles he encounters, as shown in Figure 1.

It is evident that the use of the soil type name eliminates the necessity for a detailed description of a soil profile, because the soil type name implies the possession of certain known characteristics of the various soil horizons. For example, the soil type named Chinguocousy has the following typical soils profile:

$A_1$   0 – 6″   dark grey clay loam topsoil
$A_2$   6″–12″   yellow-brown mottled clay loam
B   12″ 24″   dark brown mottled medium clay
C   24″ →   pale brown light clay till

The amount of information that can be derived from the pedological type survey for highway design purposes will be dependent upon the amount of experience the engineer has had on construction with a particular soil. The soils maps are usually prepared on the basis of shallow borings; much deeper borings are often required to obtain detailed information regarding materials to be excavated, their density, boulder content, ground water elevations, and depths of overburden. It is, therefore, necessary for highway purposes to supplement the information on the soils maps with deeper borings.

Soils maps are useful for correlating pavement performance with soil type, particularly where the "as-constructed" conditions are known. Where no previous construction history is available, particularly with respect to the depth of granular base course, it is necessary to determine the existing conditions by frequent and regular borings through the pavement.

*Example of Correlation*

Correlation of pedological soil classification with engineering classification was recently started in Peel County. Since a soil survey report and soils map were available, no actual mapping or delineation of the soil areas was required. It was necessary only to determine the engineering test results and the engineering significance of each of the horizons in the soil profile.

The first step was to determine how many samples of each horizon of each soil series would be required to give significant results. In the pedological system of soil classification, a soil type name implies that this soil has a characteristic profile and that wherever this soil type is mapped, it should have exactly the same characteristics within fairly narrow limits. This same idea can be applied to the engineering properties of each layer of the soil and the engineering test results for a given horizon of soil should fall within a narrow range. No matter where this horizon is sampled in the many different locations it may be confidently assumed that the soil is sufficiently uniform for the test values to fall within the range established.

Some agencies, in their engineering evaluation of the soil profile, do not attempt to establish a range of test values for the various horizons. The exact location of the test location is given instead, together with a detailed description of the soil profile and individual test results for that particular location. As soil series in this study were to be sampled on a county-wide basis, it was thought that the range in test results would not be too wide, and that a range would be more useful when considering a soil type in a particular application. The number of samples required to give a reasonable set of results depends upon the amount of variation within the series. Some of the factors which account for this variation are mode of deposition, topography and type of geological materials. Greater variations can be expected to occur in some soil series than in others.

Although samples were obtained from all soil horizons, the emphasis in sampling was on the parent material, because the engineer is primarily concerned with the unweathered soil as a construction material. For initial evaluation it was decided that about fifteen samples of parent material of each soil series would be obtained and a smaller number of the A and B horizons. Fewer samples have been used by other agencies; it is intended to reduce the number of samples in subsequent investigations by applying statistical methods to sampling and analysing the soil. Sample sites were located by reference to the pedological soils map, and were selected in such a way as to distribute fairly the sites in each soil series uniformly over the whole county. It was also decided to place the sampling sites near the centres of large areas of each soil series, so as not to encounter the transition zones between series. Samples were obtained from road cuts, or from test pits. Some of the samples were obtained at depths of ten feet. In all cases, depths of horizons, colour, texture, moisture conditions and any other characteristics which might be of use for identification and classification purposes were noted and recorded. Sufficient material was obtained to carry out the following laboratory tests: (1) Mechanical analysis; (2) Liquid limit; (3) Plastic limit; (4) Moisture-density relationships. In future work, tests will be made to determine the laboratory values for the California Bearing Ratio (C.B.R.).

Although results have not yet been reported, it is intended to present them in a form that can be readily used by the engineer. This will be accomplished by giving a pictorial sketch of the soil profile of each soil series, with a brief written description of each of the major horizons. This will then be followed by a table of typical engineering test values. The 90 per cent confidence limits for each test constant have been calculated, and the values will be recorded. Three classification systems will be shown. Figure 3 illustrates a typical presentation for the Oneida soil series. It is believed that it will be desirable to show also graphically the compaction and gradation curves for each soil series. The compaction curves could be used directly for field control during construction. For the compaction curves, histograms were plotted for each set of results and examined for any departures from normal, for the parent material only. A compaction curve for the Oneida series is shown in Figure 4.

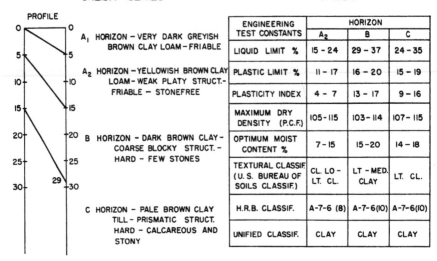

ONEIDA SERIES                                        ONEIDA

PROFILE

A₁ HORIZON – VERY DARK GREYISH
BROWN CLAY LOAM – FRIABLE

A₂ HORIZON – YELLOWISH BROWN CLAY
LOAM – WEAK PLATY STRUCT. –
FRIABLE – STONEFREE

B HORIZON – DARK BROWN CLAY –
COARSE BLOCKY STRUCT. –
HARD – FEW STONES

C HORIZON – PALE BROWN CLAY
TILL – PRISMATIC STRUCT.
HARD – CALCAREOUS AND
STONY

| ENGINEERING TEST CONSTANTS | HORIZON | | |
|---|---|---|---|
| | A₂ | B | C |
| LIQUID LIMIT % | 15 – 24 | 29 – 37 | 24 – 35 |
| PLASTIC LIMIT % | 11 – 17 | 16 – 20 | 15 – 19 |
| PLASTICITY INDEX | 4 – 7 | 13 – 17 | 9 – 16 |
| MAXIMUM DRY DENSITY (P.C.F.) | 105 – 115 | 103 – 114 | 107 – 115 |
| OPTIMUM MOIST CONTENT % | 7 – 15 | 15 – 20 | 14 – 18 |
| TEXTURAL CLASSIF (U. S. BUREAU OF SOILS CLASSIF.) | CL. LO – LT. CL. | LT – MED. CLAY | LT. CL. |
| H.R.B. CLASSIF. | A-7-6 (8) | A-7-6(10) | A-7-6(10) |
| UNIFIED CLASSIF. | CLAY | CLAY | CLAY |

FIGURE 2.

The same procedure could not be followed for the gradation curves for the soil series encountered, due to a greater variation and departure from the normal. For this reason, a graphical approach was used to determine the median and percentile ranges. This method is less accurate than using the mean and standard deviation method, but it has the advantage of being less laborious and easier to apply to non-normal distributions. Typical gradation curves for the Oneida soil series is shown in Figure 3.

*Application of Pedological Soils Information to Construction and Design*

Knowledge of the engineering properties of each soil series will aid the highway engineer considerably in making soil surveys, and in assessing the soil series as a construction material. In addition to laboratory information, it is necessary for highway purposes to know certain design and construction features associated with particular soil series. These features should be presented in tabular form in each soil survey report, dealing with the following points:

1. Is the soil adaptable for winter grading? Although it is desirable to suspend earthwork operations during the winter months to prevent the use of frozen materials in embankments, it may not be practicable to do so.

2. Normal depth to water-table. In many soil series the water-table fluctuates with the season; with a high water-table, it is evident that construction operations will be hampered.

3. Recommended location of grade line with respect to ground line. The surface and internal drainage of the soil and the soil bearing characteristics,

## GRADING CURVES - ONEIDA SERIES - PARENT MATERIAL

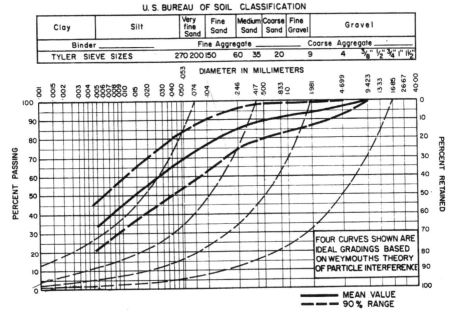

FIGURE 3.   Grading Curves for the Parent Material of the Oneida Series

particularly in areas of frost penetration, are the prime considerations when establishing a grade line.

4. Method of stabilizing slopes, particularly with reference to erosion in cuts and embankments. Slopes must be adequately protected and treated to leave a pleasing appearance.

5. Estimated percentage of boulders. On many projects, boulders are of sufficient size to require blasting and these are paid for as rock excavation; it is important to know within a fairly narrow range the quantity or volume of boulders anticipated.

6. Frost-heaving characteristics of the soil. The depth of flexible pavement over any subgrade will, to a large degree, be dependent upon the frost-heaving characteristics of the soil.

7. Suitability of the soil for fill. Although most soils are suitable for embankment construction, provided adequate compaction can be achieved, there are some soils that cannot be placed by the normal six-inch layer method.

8. Denseness of soils. In some soil series the soil is so dense that it is not possible to follow normal grading operations, and blasting must be used.

9. Are underdrains required? Underdrains are used to lower the water-table, or to intercept seepage zones. Some soil series contain seepage zones

## PROCTOR CURVE DATA ONEIDA SOILS SERIES
## PARENT MATERIAL

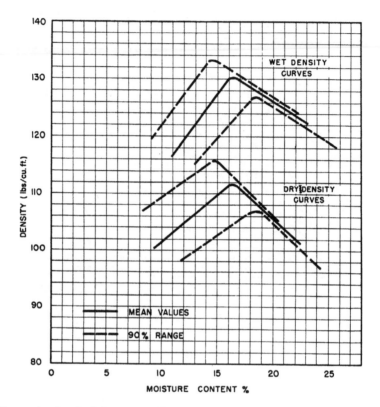

FIGURE 4.   Standard Compaction Test Results for Parent Material of the Oneida Soil Series

and underdrains must be installed to stabilize slopes or to provide good pavement performance.

10. Is the soil suitable for sub-base purposes? Sub-bases are used to provide a non-frost susceptible material over the subgrade to provide drainage and to provide strength for the pavement.

11. Is the soil suitable for granular base course? Granular base courses are processed materials, usually crushed, and placed over the sub-base; they must not be frost-susceptible and must provide stability.

12. Is this soil suitable for topsoil? Topsoil is usually stripped and reserved for replacement on slopes, borrow pits and side road developments.

It can readily be appreciated that any engineer in possession of a soils map, engineering test results and design and construction experience on each soil series, is in a good position to make adequate and competent design recommendations, with a minimum amount of field and laboratory work.

The pedological system of soil classification can be utilized for other specific highway purposes. For example, between 1948 and 1953 numerous load tests were made on several soil series throughout Southern Ontario and the bearing values in relation to thickness were associated with the soil series names. From this programme, it was possible to determine the depth of construction required over each soil type for a given loading condition. The present programme of pavement evaluation using the Benkelman Beam apparatus will be associated with soil types in a similar way.

The Ontario Department of Highways submits to contractors, tendering on Department work, a granular strip map, wherein all the known granular deposits are shown, along with quantities and suitability for a particular purpose. These granular deposits have been thoroughly investigated with power equipment and samples submitted to the laboratory for all the quality tests. The soils map together with geological maps serves to locate granular deposits for investigation and exploration purposes.

*Conclusions*

The pedological system of soil classification, as developed for agricultural purposes, is a useful tool for the highway soils engineer, because it provides him with the maximum amount of information with the minimum amount of laboratory testing. This system gives the engineer a broad concept of the soil with which he is dealing, rather than an isolated concept which is obtained by sampling and analysing soil in its disturbed state at a particular test location. The soil survey reports and maps, and laboratory test information, are not, however, sufficient within themselves to supply all the information that the engineer must have to prepare adequate designs. The engineer must still determine the precise location of the soil boundaries along the right of way, ground water elevations, organic depths in swamps and underlying soft materials, and depths of overburden, and also be aware of the construction difficulties that can arise with each soil series. When all this information has been accumulated, analysed and recorded in a suitable form, the engineer has a solid basis upon which to make adequate recommendations.

Although there are many other classification systems in use at the present time, most of these systems are derived from the results of laboratory tests, and do not take into account the environmental effects. This is of particular significance in climates where deep frost penetrations are recorded, and where wide seasonal variations in subgrade bearing values exist. The pedological system of soil classification is valuable in correlating the pavement performance with soil type.

It is evident that the soils engineer must draw his information for design purposes from many sources—from soil survey reports and soil maps, geological reports, laboratory test information, and from experience. Each source of information is a supplement to the other. When all possible information is obtained, it can best be presented on a county-wide basis to correspond with the soil survey report; it may be appended to the soils survey report and called an Engineering Chapter. The U.S. Department of Agriculture, in the preparation of their soil survey reports, are now obtaining additional detailed information in the field and engineering tests are being carried out on samples they obtain, so that the engineering chapters can be prepared with a minimum amount of additional work.

The combined pedological and engineering information can be used by many agencies:

1. To make soil reconnaissance surveys of soil and ground conditions that will aid in the selection of highways and airport locations, and in planning detailed soil surveys.

2. To locate sand and gravel deposits.

3. To correlate pavement performance with soil types, thus developing information that will be useful in designing and maintaining pavements.

4. To determine the suitability of soil units for cross-county movement of vehicles and construction equipment.

5. To make soil and land use studies that will aid in the selection and development of industrial, business and residential and recreational sites.

6. To make estimates of run-off and erosion characteristics for use in designing drainage structures and planning dams and other structures for water and soil conservation.

### REFERENCES

1. An engineering soil survey of Fayette County (1958). Kentucky Highways. U.S., Highways Research Board Bull. no. 213.
2. An evaluation of Ontario soils (1952). Ontario Dept. of Highways.
3. A soil classification for the Ontario Department of Highways (1953). Unpublished.
4. Engineering use of agricultural soil maps (1949). Highway Research Board Bull. no. 22.
5. Field manual of soil engineering, 3rd ed. (1952). Michigan State Highway Dept.
6. McLeod, Norman W. (1939). A manual of soil science applied to subgrade and base course design. Reprint from the *Canadian Engineer.*
7. Soil exploration and mapping (1950). Highway Research Board Bull. no. 28.
8. Soil Survey Reports already published (various dates). Canadian Dept. of Agriculture, Ontario Dept. of Agriculture.

# INFLUENCE OF GEOLOGY ON THE DESIGN
# AND CONSTRUCTION OF AIRPORTS

## Norman W. McLeod

EVERY MAJOR PHASE of air transport is influenced by geology. Geology affects the installation of ground facilities, the accessibility of an airport to the community it serves, and the flight patterns that aircraft must adopt when approaching or leaving an airfield, and during take-off and landing operations. Geology controls the availability and quality of airport construction materials, and may either increase or decrease the cost of providing runways, taxiways and buildings. Since the natural soil on which most airports are built has resulted in large part from geological processes, the influence of geology is also reflected in the engineering problems presented by the properties of the soil at the site.

The proper selection and development of an airport site is conditioned by three primary factors, safety, accessibility, and economy. Of these, the most important is safety. Each of these three basic considerations is affected by the geology of any proposed airfield location.

*Safety*

The geological origin of various areas has resulted in surface topographies that range from mountainous through hilly and undulating to flat. Safety considerations in air transport pertain primarily to the operation of aircraft, and it is obvious that approaching, landing at and taking off from airfields is more hazardous in mountain terrain than in flat wide-open spaces. For example, the airport at Regina, Saskatchewan, is located on the flat bed of what was once Lake Agassiz. There are no natural obstructions within many miles of the airport, and aircraft can approach it in safety from any direction.

On the other hand, the airport at Penticton, B.C., is located in a mountain valley with mountains on each side. It is apparent that aircraft can land at this airport only by approaching it along the valley itself. Runways should be so oriented that aircraft can be landed at least 95 per cent of the time with cross-wind components not exceeding fifteen miles per hour for smaller aircraft, and not exceeding twenty miles per hour for large conventional transport planes. Fortunately, regardless of the direction of prevailing winds, in a mountain valley like that at Penticton, the wind tends to blow either up or down the valley. Consequently, the runways can be oriented in the direction of the valley itself, and thereby usually provide relatively clear approaches for landing and take-off.

Geneva, Switzerland, is an example of an airport within a deep mountain recess where, after take-off, aircraft circle within the mountain enclosure to gain sufficient altitude before proceeding over the Alps. At Beirut in Lebanon, where the airport is located on a relatively narrow coastal plain at the foot of the snow-capped Lebanon Mountains, aeroplanes fly out over the Mediterranean Sea to gain sufficient height before returning to cross over these high mountains on the way to Syria, Jordan and Israel.

For safety, and for regularity of scheduled operations, an airport site should be as free as possible from ground fog, smoke, and dust. To minimize ground fog, airports should avoid low-lying areas, and should be placed preferably on higher ground. The smoke and smog generated by cities can be largely evaded by locating the airport on the windward side of the community. The application of this principle is well illustrated by the airports at Montreal, Toronto, Winnipeg, and Edmonton, for example, which have been placed on the west side of each of these cities.

To ensure greater safety during take-off and landing, the restrictions on minimum length and on the longitudinal and transverse grades of the runways are quite severe. As a minimum requirement, there must be an unobstructed line of sight from ten feet above the runway at any point, to ten feet above the runway at any other point, when these two points are separated by a minimum distance of one-half the length of the runway. Furthermore, the required minimum length of a runway is influenced by aircraft characteristics, and by the altitude, atmospheric conditions, temperature, and the over-all longitudinal gradient at the site. Consequently, it is apparent that the quantities of soil to be handled will be very much less where, as a result of its geological history, an airport site is relatively flat than where it is undulating, or where hills have to be removed and small valleys filled to satisfy the runway lengths and gradients specified.

If possible, when planning the layout for an airport, the area and shape of the site selected should eliminate any need to have the runways intersect. These intersections invariably result in an undesirable break in grade that presents a hazard to aircraft when taking off or landing. The geological features of the only site that is sometimes available may, however, make the intersection of runways unavoidable.

The geological origin of an ideal airport site will have provided a well-drained area with topography that is flat for many miles in every direction, with no hills or other natural obstructions within or between the approach zones at the ends of the runways. For airports that are to be capable of handling heavy aircraft for unlimited traffic and instrument-controlled operations, the obstruction clearance line begins at a point on the extended centre line of the runway that is 300 feet beyond and at the same elevation as the end of the runway pavement. From this point, it slopes upward at the rate of one foot vertically for every fifty feet of horizontal distance. In addition, the approach zone at each end of a runway should be free from obstructions within a trapezoidal area extending between a width of 1,200 feet at the start of the obstruction clearance line 300 feet from the end of the runway, and a width of 4,000 feet at a distance of 10,000 feet from the beginning of the obstruction clearance line. It is obviously much more difficult to comply with these requirements in hilly or mountainous terrain than on flat plains or plateaus.

Runways should be laid out so that they can be lengthened if necessary. The length of runway required for safe take-offs and landings has increased substantially during the past twenty-five years as aircraft have progressed from the small single-engine plane to the large multi-engine propellor driven aircraft. Runway length requirements have increased again with the introduction of jet aeroplanes with their much higher take-off and landing speeds. For the large jet aircraft just being introduced into commercial air transport, at full load the minimum runway length at sea level is approximately 10,000 feet, or about two miles. It is expected that the introduction of commercial supersonic flight in the relatively near future will require still longer runways, unless some acceptable system for vertical take-off and landing is perfected.

The possible need for future runway lengthening should be considered at the site selection stage. Otherwise, geological obstructions such as hills and river valleys may make runway lengthening so costly that an airport already in service, and on which large sums have been spent for ground facilities, must be abandoned, and a larger site, usually farther from the community, picked out and developed to accommodate the new larger and faster aircraft with the required degree of safety.

*Accessibility*

To encourage and facilitate the use of air transport, an airport should be located as close as possible to the community which it is intended to serve. That travel time on the ground between airport and city at the beginning and end of a trip by air frequently exceeds the time spent in flight between the two airports, is one of the most troublesome current problems in air transportation. This problem has become still more serious with the introduction of the higher speed jet aircraft.

Where the geological origin of the area has provided a flat topography, the need for minimum distance to the airport can usually be met fairly easily. The airfields at Winnipeg and Regina are good examples of the benefits of flat terrain in this respect, for they are within from 5 to 15 minutes driving time from downtown. On the other hand, minimizing the distance to the airport can become a difficult and costly problem to solve in mountainous or hilly country. The only available plateau, river valley plain, or other relatively flat area of the required size, may be many miles from the city. In this case, the requirements for easy accessibility often make it necessary to develop a nearby less favourable site at much greater cost. At Prince Rupert, B.C., for example, as a result of the mountainous mainland topography in the vicinity, the problem of accessibility was solved by locating the airport on neighbouring Digby Island, in spite of the need to make use of ferry service between the island and the city. At Hong Kong (actually

Kowloon), because of the surrounding mountains, the existing close-in airport could not provide the much longer runway with obstruction-free approaches required for the safe operation of the new large jet engine transport planes. A solution was achieved in this case by building a long runway from the present site out into the adjoining harbour, where it now roughly resembles a long pier. A similar solution was adopted for Toronto's Island Airport, where an existing runway is being lengthened by extending it out into Lake Ontario by means of dredged fill within retaining walls of sheet piling.

For Idlewild Airport at New York City, the need for ready accessibility led to the use of a nearby, low-lying shore area as the airport site, and it was enlarged to the size required by dredging the adjacent sea bottom to provide fill for the necessary area of reclaimed land. At Vancouver's Sea Island Airport, extension of the existing runways and other facilities toward the west and northwest is being carried out by the use of dredged fill over low-lying areas that are below sea level at high tide.

## Cost of Airport Runway Development

Airport runways provide the basic ground facilities required for the operation of air transport. Unlike a highway, which must usually be draped across the landscape in a generally straight line from community to community regardless of the nature of the intervening topography, drainage and soil types encountered, an airport occupies a relatively small area, and some choice usually exists concerning its exact location. Several alternate sites are ordinarily available. Each of these should be carefully studied with respect to safety of operation, accessibility to and from the community, and the cost of providing and developing the necessary ground facilities.

For a major airport, although the paved runway must have a minimum width of 150 feet, the flight strip on which the pavement is located must be carefully graded to a minimum width of 500 feet. This provides an unobstructed smooth area on either side of the pavement, on which an aircraft can be manoeuvred or brought to a stop, if it leaves the runway due to a blown tire or other difficulty. It is apparent, therefore, that unless the topography is relatively flat, the grading operations required to provide a flight strip 500 feet wide, meeting the specified rigid requirements for profile and cross-section, can become quite costly.

The excavation of solid rock is normally the most expensive unit cost item in the grading of an airport site. From the point of view of expense, this is usually followed by the unit cost of removing and replacing muskeg or swamp soils, or of other special engineering methods for treating them, while the unit cost of excavating and handling granular or cohesive soils is ordinarily much less. In so far as possible, therefore, other factors being equal, sites should be avoided that involve the excavation of large quantities of solid rock, or where the runways would have to cross appreciable areas of swamp or muskeg.

Most airports are located on ordinary soils. For these airfields, one of the most important factors influencing the cost of the needed ground facilities is the predominant type of soil that has resulted from the geological processes that have been at work in the area. If the soil consists of sand to a considerable depth, the cost of constructing the paved runways, taxiways, and aprons, will be only a fraction of the cost of these facilities at a site where the soil type is a heavy clay. At Uplands Airport at Ottawa, for example, where the soil consists of a great depth of sand, a pavement comprised of 9 inches of consolidated crushed gravel base and $3\frac{1}{2}$ inches of asphalt surfacing, constructed on this sand after it had been thoroughly compacted to a depth of 18 inches, will support the heaviest aircraft in existence, which exceed 400,000 pounds gross weight. On the other hand, at many airports where the soil consists of clay, such as, Montreal, Winnipeg, Regina, Fort St. John and Prince George, up to five or six times or more of this thickness of granular base and surfacing (depending on the characteristics of the clay soil) would be required to sustain landing, take-off, and taxiing operations by the world's heaviest planes. (The asphalt surface would still be limited to $3\frac{1}{2}$ to 4 inches.) Consequently, there is an important economic incentive to locate an airport on sandy rather than on a clay soil, whenever this choice is available.

Large quantities of aggregate materials such as sands, gravels, quarried and crushed stone are needed for the paved areas at an airport. Depending upon the past geological history of any given region, these construction materials may be available locally, or they may have to be brought in from considerable distances. Long distance haulage adds substantially to the cost of these materials, and to the cost of the completed airport. In southwestern Manitoba, for example, there is a large area of sand, but no gravel deposits occur for many miles. At the airport at Sudbury, Ontario, on the other hand, gravel in large quantities exists on and adjacent to the site.

Because of the repeated glaciation of the northern half of this continent, the most common sources of aggregate materials for airport construction in Canada are certain common glacial features such as moraines, eskers, kames, ancient beaches, stream terraces, river beds, and occasionally wind-deposited sand dunes. In some parts of Canada, particularly southwestern Ontario, and in the Montreal area, natural deposits of gravel have become so seriously depleted that large stone quarries have been opened to supply the aggregates required.

The location of natural aggregate deposits has been facilitated by the development of two relatively new techniques, electrical resistivity, and aerial photography. Clay and silt soils ordinarily have comparatively low electrical resistance, while that of sand, gravel and solid rock is usually high. By passing direct current through two electrodes some distance apart, and measuring the voltage drop between two intermediate electrodes, sand and gravel deposits can be located. Aerial photographs register detailed differences in topography, drainage, vegetation, etc., due to the geological origin of the region that has been photographed. By means of stereoscopic examination, a technician who is skilled in the interpretation of aerial photographs, can pick out probable locations of aggregate deposits from the photographs themselves.

## Some Items of Special Interest

1. In the area in which the airport at Winnipeg is located, the soil consists of a heavy gumbo clay. At a depth of from about one to two feet below the surface, however, there is a layer of fine sand or silt that varies from a fraction of an inch to several inches in thickness, with pockets that may be several feet deep. Below this layer, the soil again consists of clay to considerable depth. This layer of fine sand or silt is a source of serious frost heaving or frost boils. Consequently, during the grading operations for the runways at the Winnipeg airport, borings were made to check for the existence of this layer, and it was removed wherever it occurred.

2. At Digby Island on which the airport for Prince Rupert was built, the natural terrain at the site consisted of rock outcrops and muskeg. The muskeg had to be excavated to its full depth and replaced to the required elevation with granular material, and the rock outcrops were blasted to grade. The quantities handled were 1,319,128 cu. yds. of muskeg, and 979,380 cu. yds. of solid rock.

3. Due to the geological history of the region, deposits of iron and coal occur at Sydney, Nova Scotia. Large quantities of blast furnace slag have accumulated from the production of iron by Dominion Steel and Coal Corporation, Ltd. Since it was readily available and of satisfactory quality, this blast furnace slag was employed as the base course aggregate for the runways at the Sydney Airport.

4. Because of the local geology and climate, the new airport being constructed at Inuvik, near the mouth of the Mackenzie River, is located on permafrost. Experience has indicated the need for maintaining the natural ground in its permanently frozen condition when engineering structures are placed on permafrost. The existing cover of muskeg is such an excellent insulator that it should be disturbed as little as possible. During the construction of a smooth airport runway over the uneven tundra terrain, some disturbance of the moss cover is inevitable, even if it is only compressed by the weight of the overlying layer of applied granular material. The minimum thickness of the granular layer must be sufficient to compensate for this loss of insulation efficiency by the moss, and to ensure freezing temperatures throughout the year at the surface of the permafrost layer just under the moss. For the first runway for the new airport at Inuvik, a depth of eight feet of granular material was placed with minimum disturbance to the natural ground. Temperature measuring devices placed at the bottom of this fill, and throughout the embankment, have indicated so far that this 96-inch layer of superimposed granular material is just sufficient to avoid a temperature higher than 32°F at the underlying natural ground level.

*Conclusion*

As this paper has shown, geology has an important influence on many aspects of air transport. It affects the three basic factors to be considered in connection with the selection and development of an airport site, safety, accessibility, and economy. Safety is influenced by the existence or absence of natural obstructions of geological origin which control the flight patterns that aircraft must employ when approaching or leaving an airport. The geological history of the region affects an airport's accessibility, which is measured chiefly in terms of driving time between an airfield and the community it serves. Finally, there are important determining local features such as topography, drainage, nature of the soil, presence or absence of solid rock, swamp, muskeg, or permafrost, and availability and quality of aggregate materials. These have resulted from the geological processes that have been at work in the area, and they largely determine the cost of constructing the runways, taxiways, buildings and other facilities that are required for the movement of aircraft on the ground, and for the handling of passengers, baggage, and air cargo.

# ENGINEERING STUDIES OF LEDA CLAY

## C. B. Crawford

LEDA CLAY WAS DEPOSITED in an inland sea called the Champlain Sea, which extended up the Ottawa river valley to Pembroke, up the St. Lawrence to Kingston and down the St. Lawrence valley to the ocean. Over much of this area the clay is covered with a mantle of sand. In his classic *Geology of Canada* (1863) Sir William Logan noted that "Dr. J. W. Dawson, who has carefully studied these deposits in Canada, distinguishes the lower as the Leda clay from one of its characteristic shells; and the upper, for a similar reason, as the Saxicava sand." This was apparently the first recorded use of the term "Leda" clay. Later Dawson reported his extensive investigations of the clay, noting its unusual properties and the fact that it had been derived from local shales. He opened the long-discussed geological question of why the marine clays are not found in the Lake Ontario basin (J. W. Dawson, 1893).

In recent years considerable advance has been made in the geological understanding of Leda clay. In particular, the advent of radiocarbon dating and palynological studies, coupled with more detailed field exploration, has established the Champlain Sea as late glacial rather than post-glacial. The Champlain Sea is now considered to be contemporaneous with the Two Creeks interval of late Wisconsin age and to have been a single marine invasion followed by estuarine and lacustrine conditions due to isostatic uplift (Terasmae, 1959; Gadd, 1960). The marine period is known to have been influenced greatly by the inflow of fresh water, creating brackish water conditions (Gadd, 1957). Even the name "Leda" has been questioned on the basis of precedence of generic names (Gadd, 1960), but it has the advantage of general acceptance and familiarity.

Mineralogical studies have shown the clay to contain substantial amounts of illite and chlorite with lesser amounts of inert materials such as quartz and feldspars (Lambe and Martin, 1955). Mineral composition is known to vary with size fraction. In general the silt fraction ($>$ 0.002 mm) consists of rock flour derived from the local Precambrian rocks although this inert material is known to extend into the finer fractions. The clay fraction is predominantly composed of illite and chlorite with occasional traces of montmorillonite (Karrow, 1960; Forman and Brydon, 1960). According to a recent study, "typical Quaternary Clays of New England and Eastern Canada resemble the shales of the area in mineral composition; they are composed chiefly of potash clay minerals, or hydrous micas, and chlorite or vermiculite" (Allen and Johns, 1960).

*Engineering Investigations*

It had been common engineering practice until recently to establish the extent of Leda clay by wash boring or by some crude sampling technique. The more important engineering structures were usually founded on piles. Significant laboratory testing was unknown and field testing was usually confined to hand sounding for the purpose of extending previous experience.

With the development of refined soil sampling and testing it was found possible, and desirable, to apply the techniques of modern soil engineering to this sensitive clay; to this end a great deal of subsurface investigation has been done. Although little of this work has been published, it is not surprising that one of the first modern studies to be made available was concerned with the classic earth flow near St. Thuribe, Quebec, which occurred more than sixty years ago (Peck, Ireland and Fry, 1951). This earth flow was, at the time of occurrence, subjected to a comprehensive geological study which included soil tests such as density, water content and grain size. Further, the report of this study included a rather modern explanation of the failure as follows: "It appears probable that in this particular instance the silty-clay, surcharged with water, stood in a condition of unstable equilibrium, retaining its solidity merely by virtue of its unbroken molecular texture, and that at the moment in which it became subject to internal movement this texture gave way and it lapsed into a nearly liquid mass, the particles re-arranging themselves with some freedom in the water previously locked up in its pores" (G. M. Dawson, 1899).

From published information, and recent test results by the N.R.C. Division of Building Research, the engineering properties of Leda clay covering a broad region are summarized in Table I. Liberties have been taken in establishing average values but, where appropriate, maximum and minimum recorded values are quoted in parentheses. The table illustrates a noticeable trend. With increasing distance inland from the ocean, the clay size content, plasticity and natural water content increase. Liquidity index and sensitivity vary between wide limits in specific locations. The clay is relatively "inactive" in all regions. The salt content of the pore water is variable and puzzling. After a considerable number of tests on extracted pore water in the Ottawa area, only one location revealed a substantial salt content in the pore water (up to 15 grams/litre). Less than one mile away another boring at about the same elevation revealed less than 2 grams/litre.

Leda clay commonly exhibits a black mottling in fresh samples from below the water table. The mottling disappears on exposure to air. On the suggestion of Dr. I. Rosenqvist, several of these samples were examined by Dr. N. E. Gibbons of the Division of Applied Biology at the National Research Council and were found to contain colonies of anaerobic bacteria which were responsible for the mottling. The only location in the Ottawa vicinity known to be free of black mottling is the area of high salt concentration.

The undrained shear strength of Leda clay varies widely and is directly

TABLE I
ENGINEERING PROPERTIES OF LEDA CLAY

| Location | Approx. Elevation (ft.) | Natural Water Content (w/c %) | Liquid Limit (LL %) | Plastic Limit (PL %) | Liquidity Index (LI) | Sensitivity (St) | Clay Size (%) | Activity | Salt Content (gram/litre) | Undrained Shear Strength (tons/sq. ft.) | Reference |
|---|---|---|---|---|---|---|---|---|---|---|---|
| QNSLR | +260 −125 | 38 (20–70) | 34 | 19 | 1.3 | | (43–65) | 0.5 | (0.5–5) | 1.1 (0.5–2) | Pryer and Woods (1959) |
| St. Thuribe | +175 | 44 | 33 | 21 | 1.9 | 8 | 36 | 0.3 | | 0.8 | Peck, Ireland and Fry, (1951) |
| Nicolet (Upper) +50 20 | | 65 | 55 | 23 | | 9 | 73 | 0.4 | 0.5 | 0.3 | Hurtubise and Rochette (1956) and Crawford and Eden (1960) |
| Nicolet (Lower) +20 −30 | | 53 | 42 | 23 | | 10 | 50 | 0.5 | 9 | 0.6 | |
| Beauharnois | +140 100 | 68 (50–80) | 59 (44–68) | 24 (22–26) | 1.3 | 12 (7–23) | 64 (47–72) | 0.5 | 0.4 | 0.5 | Eden and Hamilton (1957) |
| Massena | +215 200 | 60 (50–75) | 50 (35–60) | 27 (20–35) | 1.5 | (7–∞) | 45 (30–55) | 0.5 | (0.6–2) | 0.3 (0.1–0.6) | Burke and Davis (1957) Bazett, Adams and Matyas (1960) DBR unpublished |
| Hawkesbury | +225 | 80 (61–90) | 64 (53–72) | 26 (25–28) | 1.4 | 27 (9–∞) | 74 (65–88) | 0.5 | 0.3 | 0.4 | Eden and Hamilton (1957) |
| Ottawa (Upper) +310 200 | | 80 (60–90) | 60 (50–75) | 25 (20–30) | 1.6 | 24 (6–∞) | 70 (55–80) | 0.4 | 1.0 | 0.7 (0.3–1.0) | Eden and Crawford (1957) and DBR (unpublished) |
| Ottawa (Middle) +200 170 | | 45 (35–60) | 32 (20–50) | 21 (14–27) | 3.0 | — | 64 (60–75) | 0.4 | 1.5 | 0.7 (0.6–1.0) | |
| Ottawa (Lower) +170 100 | | 65 (50–70) | 65 (60–75) | 28 (27–30) | 1.2 | (5–∞) | 73 (60–75) | 0.6 | (0.5–15) | 1.2 (0.5–1.5) | |

NOTE: LI = (w/c − PL)/(LL − PL).    Activity = (LL − PL)/(% clay size).    Clay size = finer than 0.002 mm.

related to the preconsolidation pressure of the natural clay (Eden and Crawford, 1957; Eden, 1959). With few exceptions the apparent cohesion ranges from 0.3 to 0.4 times the preconsolidation pressure.

Leda clay is often but not always homogeneous in structure. Occasionally it is extremely stratified. Gadd (1960) draws attention to the banding that ". . . but for the fossils, could readily be mistaken for glacial varves." Pryer and Woods (1959) pay special attention to the banding and note the wide variations of water contents and Atterberg limits. Failure to take this into account, they point out, results in unrealistic values of liquidity index.

### Landslides

The property of Leda clay that distinguishes it from other soils is its sensitivity (the ratio of its undisturbed to its remoulded strength). In extreme cases the clay becomes a viscous liquid when remoulded. This is illustrated in Figure 1 where a slice has been removed from an undisturbed specimen, remoulded, and deposited beside the specimen. In this respect,

FIGURE 1.    Illustration of sensitivity of Leda clay

it is similar to few of the known soils of the world. Only the Scandinavian marine clays have a similar loss of strength on remoulding. Correspondingly, only from the marine clay regions of Scandinavia and of eastern Canada have come reports of earth flows, or flow slides as they are sometimes called. A map showing major landslides in the St. Lawrence and Ottawa valleys has been published (Hurtubise, Gadd and Meyerhof, 1957). The most recent serious earth flow destroyed part of the village of Nicolet, Quebec, in 1955.

Earth flows have a characteristic mode of failure and final shape. They appear to be initiated by a rather small bank failure which then retrogresses

## TABLE II
### EARTH FLOWS AND LANDSLIPS IN LEDA CLAY

| No. | Date | Name | River | Acres | cu. yd. | Victims | References |
|---|---|---|---|---|---|---|---|
| 1 | Dec. 7, 1955 | Hawkesbury | | 12 | 500,000 | | Eden (1957) |
| 2 | Nov. 12, 1955 | Nicolet | Nicolet | 6 | 250,000 | 3 | Hurtubise and Rochette (1956) Crawford and Eden (1960) |
| 3 | Aug. 3 } 1951 Aug. 6 | Rimouski | Rimouski | 25 | 1,000,000 | | Meyerhof (1954) |
| 4 | May 18, 1945 | St. Louis | Yamaska | 7 | | | Béland (1956) |
| 5 | Sept. 9, 1938 | Ste. Geneviève | Batiscan | 15 | | | Béland (1956) |
| 6 | July 24, 1935 | St. Vallier | Des Mères | 15 | | | Béland (1956) |
| 7 | Fall, 1924 | Matteau Farm | St. Maurice | 10 | | | Hodgson (1927) |
| 8 | Apr. 6, 1908 | Notre-Dame de la Salette | Lièvre | 6 | | 33 | Ells (1908) |
| 9 | Oct. 11, 1903 | Poupore | Lièvre | 100 | | | Ells (1904) |
| 10 | May 7, 1898 | St. Thuribe | Blanche | 86 | 3,500,000 | 1 | G. M. Dawson (1899) |
| 11 | Sept. 21, 1895 | St. Luc de Vincenne | Champlain | 5 | | 5 | Hodgson (1927) |
| 12 | Apr. 27, 1894 | St. Alban | Ste. Anne | 1,600 | 25,000,000 | 4 | LaFlamme (1894) |
| 13 | Apr. 4, 1840 | Maskinongé | Maskinongé | 84 | | | Logan (1842) |
| 14 | — | Green Creek | Ottawa | 30 | 2,000,000 | | |
| 15 | — | Azatika Creek | Ottawa | 10,000 | | | |

in a series of slides in which part of the soil is completely remoulded and flows out of the small opening in the bank carrying large chunks of relatively undisturbed soil with it. The resulting crater is usually pear-shaped. Less frequently the entire bank will be removed with the slide and the material will flow away down the most convenient channel. Table II lists several of the documented landslides that have occurred in Leda clay.

There is circumstantial evidence to suggest that potential earth flows are set up by geologic processes such as climatic cycles, gradual erosion, *in situ* changes in soil properties and groundwater changes, and are finally triggered by a chance occurrence such as rapid erosion, vibration, or construction work.

The potential instability begins with the natural downcutting of a stream. Heavy precipitation loads the unstable zone and increases the stresses in the subsoil. Changes in soil properties, such as leaching of salt from the pore water, may increase sensitivity and decrease shearing strength (Bjerrum, 1954). Finally an incipient failure condition is reached naturally, and a chance occurrence sets off the slide.

One of the best examples of an earth flow to be revealed in the scar-studded banks and terraces of the St. Lawrence and Ottawa rivers is shown in Figure 2. Here two million cubic yards of sensitive clay have covered seventy acres of bottom land to a depth of fifteen to twenty feet leaving a 30-acre crater in the 90-foot high bank. The flow, at the eastern limits of Ottawa, occurred before recorded history and after the withdrawal of the Ottawa river to its present channel.

A much larger earth flow, along the Quyon river, just west of Ottawa, was reported by Wilson (1924). Similar to the Quyon flow is the large earth flow shown in Figure 3. This flow, covering about fifteen square miles in Alfred Township, Prescott County, Ontario, was discovered during a field investigation in the area. The most careful sampling techniques had been yielding apparently disturbed material, and from surface observations it was suspected that the area being sampled might have been within an old earth flow. Subsequent examination of air photos revealed the great extent of the flow. It is apparent, therefore, that earth flows in Leda clay may have much engineering significance.

Not all landslides in the Leda clay are of the flow-type. Several small landslides occurred in the marine clay along the Moisie River Valley during construction of the Quebec North Shore and Labrador Railway. One slide of about three thousand yards in a 75-foot high bank failed along a nearly perfect circular arc a few months after the cutting of the slope (Pryer and Woods, 1959). A few yards away a typical flow-type failure occurred when the roof of a rock tunnel failed allowing sixty-thousand yards of liquefied clay to flow through the hole (Monaghan, 1954). This is the characteristic pattern; normal circular type slope failures in the same vicinity as earth flows.

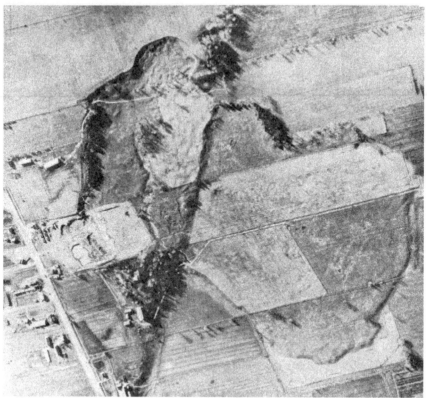

FIGURE 2. Earth flow near Ottawa, Ontario (City of Ottawa. Photo by A. E. Simpson Ltd., Montreal).

## Shear Strength

The most difficult soil engineering problems involve the shearing resistance of soil. Economy dictates low safety factors, often less than 1.5. In view of the natural variations in soil properties the occasional failure of cuts, fills and natural slopes must be accepted. Small failures, if anticipated, are not usually serious, but in the case of extra-sensitive soils small failures often lead to catastrophic failures.

Rigorous analysis of the stability of natural slopes in sensitive clays has not yet been particularly successful. The analysis is usually complicated by a surface layer of desiccated soil having a wide range of strength values which cannot be properly assessed. Further, the prospect of attempting to analyse the miles and miles of natural slopes in earth-flow territory is staggering. Even the analysis of failures is complicated by the drastic changes which occur during the movement.

In view of the complete liquefaction that sometimes occurs when Leda

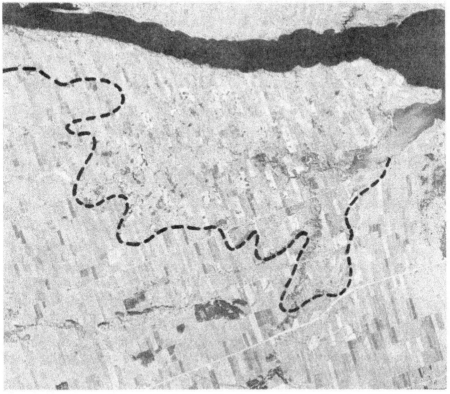

FIGURE 3.  Earth flow, Alfred Township, Prescott County, Ontario (RCAF photo A14526–49).

clay is stressed to failure, it is not convenient to initiate failures, and strength research is therefore largely confined to the laboratory. Because pore pressures vary and water moves in soil due to stress application, the total stresses applied during testing may bear little relation to effective stresses (that is, the so-called grain-to-grain stress). Since it is known that strength and deformation depend on changes in effective stresses, most of the current laboratory research is done in a triaxial shear apparatus so that pore pressure changes can be measured while the specimen is brought to failure under a variety of stress conditions.

Laboratory specimens have been subjected to various incremental tests and various rates of strain (Crawford, 1959). Pore pressure changes under constant deviator stress have been measured. The amount and rate of strain have been shown to have an appreciable effect on shear strength. Significant movement of pore water away from the failure plane has been observed. It is deduced that under the application of shear stresses the delicate structure of the clay will collapse and transfer stresses to the pore water. When these pore water stresses are sufficiently developed, and effective stresses are correspondingly reduced along a continuous path, a sudden shear failure can occur. This is apparently the mechanism of failure in the field.

It is not possible, however, to create a sudden failure condition in the usual controlled-strain triaxial test because pore water tends to move away from the plane of maximum stress causing an increase in strength and a shift in the location of the critical failure plane. Sudden failure can be accomplished, however, in controlled stress laboratory tests. In a series of controlled stress tests carried out by K. N. Burn of the N.R.C. Division of Building Research, it was observed that specimens subjected to a constant deviator stress would strain slowly for periods ranging from one hour to five hours, depending on the applied stress, and would then fail in less than five minutes. In general the maximum deviator stress reached by controlled straining agreed with that required for failure by controlled stress application, if the time to failure was the same in each case.

Typical Leda clay specimens fail at a strain of 1 or 2 per cent. Maximum deviator stress is at least reached within this strain range. Continued straining causes only slight reduction in the deviator stress but considerable increase in pore water pressure. As a result, the effective stresses are reduced without appreciable reduction in shearing resistance so that, in terms of effective stresses, the shear strength increases with increasing strain (Crawford, 1959). It has been suggested (Kenney, 1959) that shear strength parameters are not fully "mobilized" at 1 or 2 per cent strain and that the higher shearing resistance measured at greater strains is the true value. This concept introduces a controversial criterion of failure and permits a wide range of interpretation of shear strength parameters in terms of effective stresses.

There is obviously a great difference between failure in the laboratory and failure *in situ*. There may also be a substantial difference between pore water pressures measured in the triaxial test and those that occur on the plane of failure at the instant of failure. These differences limit the usefulness of laboratory studies of the sensitive clay and must ultimately be resolved by field studies.

Fortunately, since Leda clay is only slightly overconsolidated, most soil engineering analyses can be made satisfactorily in terms of total stresses, using shear strength values obtained by unconfined compression or vane test. This is true only because in most practical situations the strength of a normally consolidated clay increases with time after application of stresses. When this is not the case, it is necessary to know the changes in effective stresses which occur with time. This latter qualification to total stress analyses may be quite important in dealing with the desiccated surface crust. It should be noted that stress-deformation problems in soils which involve long periods of time cannot be evaluated properly in terms of total stresses, but until more confidence can be shown in the determination of effective stress parameters at failure in sensitive clay, the total stress method offers some advantages.

A comparison of many tests indicates that the strength determined *in situ* by a vane test generally exceeds that given by laboratory unconfined-compression tests, the lower laboratory values being only one-third to one-half of the field measurements, while the maximum laboratory values agree reasonably well with field determinations (see, for example, Hurtubise and Rochette, 1956; Eden and Hamilton, 1957). Specimens from the upper, stiff-fissured zone often fail due to splitting in unconfined compression; unconsolidated-undrained triaxial tests are preferred in these cases. Skin friction on the vane torque rod has been found to have an appreciable effect on test results, but a limited investigation showed little effect of vane area ratio (Eden and Hamilton, 1957).

Special studies of the sensitivity and thixotropy of Leda clay are in progress. The possible variations in measured sensitivity depending on test method have been indicated (Eden and Hamilton, 1957). The most sensitive specimens are particularly susceptible to the method used for testing. Thixotropic regain of strength after remoulding, on specimens stored for periods of more than a year, has shown a significant increase with time.

Preliminary observations of the effect of an electric current on adhesion of the clay to metal have been made. In the course of using a special sounding device, it became stuck in the clay due to thixotropic hardening. A pull of 13 tons was applied to the 70-foot deep string of drill rods without success. After an application of 36-volts direct current, however, delivering about 12 amps for three or four minutes on the stuck drill rods as the cathode, the rods moved freely under their own weight. The technique has obvious practical possibilities.

*Settlement Problems*

Leda clay generally has a high proportion of void space to solid material; often less than one-third of the volume of a saturated specimen of soil will be solid material. The solid particles which form the soil structure are plate-shaped, probably equivalent in dimension to razor blades and arranged predominantly in an edge-to-face contact sequence (Rosenqvist, 1959). Under externally applied loads the soil structure is compressed and some of the void water is squeezed out. Correspondingly if water is extracted from the soil by drying, the soil structure shrinks due to pore water suction forces. The resulting magnitude of volume change is illustrated in Figure 4 by two similar specimens of soil, each of which was trimmed to the same volume and one of which was allowed to dry naturally from 60 per cent to 30 per cent water content.

This characteristic "open structure" and high compressibility results in two distinct types of settlement failure. On the one hand a deep-seated consolidation settlement occurs when the soil is overloaded, and on the other hand lightly loaded, shallow foundations settle due to shrinkage of the soil

FIGURE 4.    Shrinkage of Leda clay specimens
Specimen A — w/c = 60%
Specimen B — w/c = 30%

by drying. Owing to consolidation, many of the large churches along the lower St. Lawrence river, for example, have experienced many inches of settlement. The National Museum in Ottawa is a classic case of consolidation settlement. The extent of the shrinkage problem is not so well known. In Ottawa, however, settlements of shallow foundations have exceeded 1 foot due to this cause (Bozozuk, 1957).

### Shrinkage and Swelling Studies

Laboratory investigations have shown that the shrinkage of Leda clay is only partly reversible. Figure 5 shows test results on three soil specimens, which were partially or wholly dried from an initial water content of about 77 per cent and then allowed to swell with complete water availability (Warkentin and Bozozuk, 1960). The substantial volume reduction after a complete dry-wet cycle indicates an irreversible structural change. The vertical and horizontal shrinkage of undisturbed specimens were about equal.

To determine the natural seasonal variation in soil water content, borings were made twice a month from April 1955 through March 1956, at a selected well-drained grassy area near the Building Research Centre, Ottawa. Samples for water content analysis were obtained whenever possible at the surface, at 3-inch increments to a depth of 2 feet, and at 6-inch increments from 2 to 8 feet. Test results were averaged by computing the mean value of three measurements obtained just above, below and at depths of 6, 12, 24, 48, and 90 inches. The maximum water content variation so obtained is shown in Figure 6. Very little water content change is noted

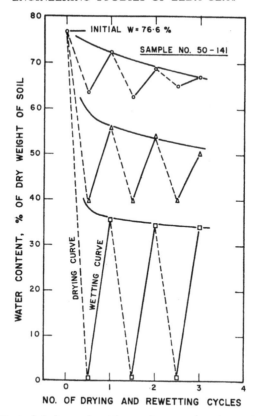

FIGURE 5.    Effect of drying and wetting cycles on end water content of Leda clay
(after Warkentin & Bozozuk, 1960)

during the entire year below a depth of 3 feet. The measured variation of
3 or 4 per cent could easily be accounted for by non-homogeneity. At the
one-foot depth the annual range in water content is from 15 per cent to
35 per cent, while at the surface it appears to vary from less than 10 per
cent to more than 60 per cent. Much of the high water content in the upper
foot of soil occurs during the winter months and is therefore attributed to
ice lensing due to frost action.

Near the site where seasonal water content was studied, and towards a
row of large elm trees, ground movement gauges were installed at various
depths and distances from the trees. Periodic measurements confirmed the
suspected influence of trees on the desiccation and shrinkage of the clay
subsoil (Bozozuk and Burn, 1960). The magnitude of ground movements
in relation to the trees during the exceptionally dry and hot summer of 1955
is shown in Figure 7.

Laboratory studies and field measurements have shown that vertical
ground movements of considerable magnitude can occur in Leda clay due

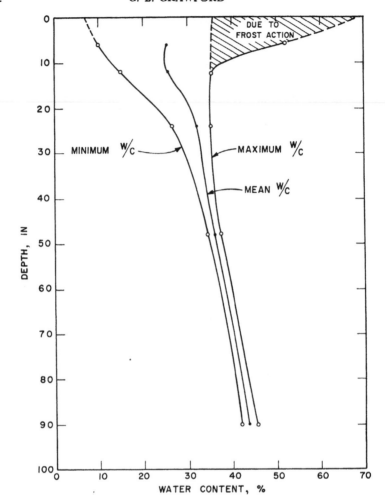

FIGURE 6.   Seasonal variation in soil water content of Leda clay

to seasonal drying. It is reasoned from these observations that the absolute settlement of shallow foundations is cumulative with the greatest increments occurring during exceptionally dry seasons. Movements increase sharply near trees and the presence of trees is therefore often the cause of major differential movements. Further, it has been noticed that movements are most serious where a mantle of loose sand occurs above foundation level, a situation that apparently facilitates deep climatic influences and drives tree roots deeper into the clay. Field experiences have demonstrated partial reversibility of the shrinking. Building cracks can often be partly closed by artificial wetting (see for instance Legget, 1954), but complete reversibility has not been achieved in the field or laboratory.

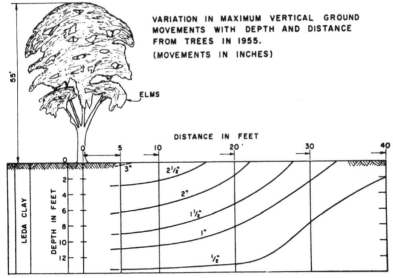

FIGURE 7.    Measured ground movements near a row of trees

In general, it can be concluded that the upper ten feet of Leda clay, if at a water content exceeding 40 per cent, are particularly susceptible to shrinkage; in shallow foundation design special efforts should be made to prevent differential drying.

### Consolidation Studies

Leda clay exhibits compression characteristics common to all sensitive clays. Deflection is slight over the recompression range, with a sudden break at the preconsolidation load. The compression index is usually greater than 1 and frequently it is greater than 2.

In the investigation of the settlement of the National Museum Building in Ottawa (Crawford, 1953), it was found that computed settlements ranged up to about four times as great as measured settlements. The complexities of the foundations for the building created serious difficulties in computing stresses in the ground, and certain assumptions had to be made regarding rigidity of the building. It was also suspected that the theoretical evaluation of the soil compressibility was contributing to the disagreement. In particular, laboratory evaluation of the preconsolidation pressure was suspect.

To evaluate preconsolidation pressure *in situ* it was necessary to find a construction project involving simple loading of a thick deposit of Leda clay and to measure, in the field, compression of the subsoil under known loads. Such a project became available, through the co-operation of the Ontario Department of Highways, in the form of a 23-foot high earth fill over a 100-foot deposit of clay. Borings were made, samples were obtained, and special remote reading instruments were developed for measuring subsoil compression (Burn, 1959). Preliminary measurements have confirmed the suspected underestimation of preconsolidation pressure.

In the Ottawa region the measured preconsolidation pressure can be correlated with absolute elevation (Eden and Crawford, 1957). This engineering evidence provides a physical basis for geological interpretation of the original Champlain sea bed elevation in the area and supports the opinion that surface erosion has left the local Leda clay slightly preconsolidated. Further, it provides a method of evaluating test results. Figure 8 shows several series of test results ranging from elevation 80 to 260 above sea level.

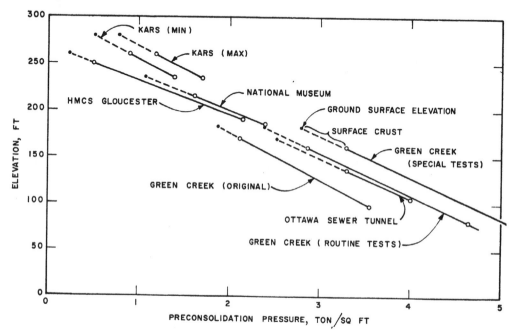

FIGURE 8.   Relation between preconsolidation pressure and elevation

Attention is invited to the three curves labelled "Green Creek," which is the site of the earth fill study mentioned above. The lower curve was based on only four consolidation tests from preliminary borings (Figure 7, Eden and Crawford, 1957). The middle curve was based on a series of routine tests from commercial samples. The upper curve was obtained from more refined tests done on very carefully obtained samples, using a thin-walled fixed piston sampler with an area-ratio of only ten per cent. All samples were taken at the same location but the quality of sampling and testing has had a decided effect on the final result. As more results become available, individual test values may be evaluated with confidence using this general relationship. From the results shown, it can be concluded that refined sampling and testing result in considerably higher estimations of preconsolidation pressure than would otherwise be obtained.

Research on borehole samples has the great disadvantage that properties change with depth and the effects of variations in test procedure are masked by this variation. Block samples of Leda clay, carefully cut from a deep excavation, were therefore obtained for further study. These studies (Hamilton and Crawford, 1959) revealed difficulties in interpreting the "most probable preconsolidation pressure" of the Leda clay from routine incremental loading tests. They further indicated a substantial effect upon the test results by gas in the specimen (as a result of unloading due to sampling), stress variations in the specimen during test, and they suggested that secondary compression (due to structural collapse) was an important part of the measured compression.

Subsequent laboratory tests have confirmed these suggestions. In addition, pore pressure measurements have shown that almost all of the measured deflection at pressures near the preconsolidation pressure is secondary compression; beyond the preconsolidation pressure about one-third of the deflection is of a secondary nature. If secondary compression is more prominent in the laboratory test than it is *in situ*, this will be a major source of error in interpreting laboratory results.

There is some evidence that secondary compression is not prominent *in situ*. The National Museum, for example, is not thought to have experienced significant settlement since 1915, only five years after its completion. If the settlements were of a secondary nature, movements would still be occurring. A second earth fill has been instrumented to include not only measurements of subsoil compression but also pore pressure increase and dissipation, and lateral subsoil movement due to loading. This test fill is expected to give valuable information on secondary compression.

### Conclusions

Combined geological and engineering effort has removed much of the mystery surrounding the unusual properties of Leda clay. Greatly improved clay mineralogy equipment and techniques together with the increasing application of soil physics is expected to reveal more fundamental properties of the clay.

Only combined laboratory and field investigations will disclose the true strength and deformation properties of the material. It is in this direction that current research of the Division of Building Research is proceeding.

#### ACKNOWLEDGEMENTS

This paper is a contribution of the Division of Building Research of the National Research Council of Canada. It is published with the approval of the Director of the Division, Mr. R. F. Legget, whose interest and encouragement is especially appreciated. Work of the author's colleagues in the

Soil Mechanics Section of the Division of Building Research has been quoted freely and is gratefully acknowledged.

## REFERENCES

1. ALLEN, V. T. and JOHNS, W. D. (1960). Clays and clay minerals of New England and Eastern Canada. Bull. Geol. Soc. Amer. *71*: 75–86.
2. BARACOS, A. and BOZOZUK, M. (1957). Seasonal movements in some Canadian clays. Proceedings, Fourth International Conference on Soil Mechanics and Foundation Engineering, London, vol. 1, Div. 3, 264–8.
3. BAZETT, D. J., ADAMS, J. I. and MATYAS, E. L. (1960). An investigation of slides in a test trench excavated in fissured sensitive marine clay. Preprint of submission to the Fifth International Conference on Soil Mechanics and Foundation Engineering, Paris, 1961.
4. BÉLAND, J. (1956). Nicolet Landslide, November 1955. Proc. Geol. Assoc. Canada, *8*, pt. 1: 143–56.
5. BJERRUM, L. (1954). Geotechnical properties of Norwegian marine clays. Géotechnique, vol. IV, no. 2, 49–70.
6. BOZOZUK, M. (1956). Settlement studies of light structures in Ottawa. Proceedings, Tenth Canadian Soil Mechanics Conference, N.R.C., Associate Committee on Soil and Snow Mechanics, Tech. Mem. no. 46, 53–4.
7. BOZOZUK, M. and BURN, K. N. (1960). Vertical ground movements near elm trees. Géotechnique, vol. X, no. 1, 19–31.
8. BURKE, H. and DAVIS, W. (1957). Physical properties of marine clay and their effect on the Grass River Lock excavation. Proceedings, Fourth International Conference on Soil Mechanics and Foundation Engineering, vol. 2, 301.
9. BURN, K. N. (1959). Instrumentation for a consolidation study of a clay deposit beneath an embankment. Géotechnique, vol. IX, no. 3, 136–46.
10. CRAWFORD, C. B. (1953). Settlement studies on the National Museum Building, Ottawa, Canada. Proceedings, Third International Conference on Soil Mechanics and Foundation Engineering, Switzerland, vol. 1, 338–45.
11. ———— (1959). The influence of rate of strain on effective stresses in sensitive clay. Presented to the 62nd Annual Meeting, American Society for Testing Materials.
12. ———— and EDEN, W. J. (1960). The Nicolet Landslide of November 1955 (in press).
13. DAWSON, G. M. (1899). Remarkable landslip in Portneuf County, Quebec. Bull. Geol. Soc. Amer. *10*: 484–90.
14. DAWSON, SIR WILLIAM (1893). The Canadian ice age. Published by Dawson, Montreal, pp. 52–61.
15. EDEN, W. J. (1956). The Hawkesbury landslide. Proceedings, Tenth Canadian Soil Mechanics Conference, N.R.C., Associate Committee on Soil and Snow Mechanics, Tech. Mem. no. 46, 14–22.
16. ———— and CRAWFORD, C. B. (1957). Geotechnical properties of Leda clay in the Ottawa area. Proceedings. Fourth International Conference on Soil Mechanics and Foundation Engineering, London, vol. 1, 22–7.
17. ———— and HAMILTON, J. J. (1957). The use of a field vane apparatus in sensitive clay. ASTM Special Technical Publication no. 193, 41–53.
18. ———— (1959). Discussion to "Investigation of banded sediments along St. Lawrence North Shore in Quebec." ASTM Special Publication no. 239, 70–3.
19. ELLS, R. W. (1904). The recent landslide on the Lièvre River, P.Q. Geol. Surv. Canada, Ann. Rept. XV, 136A–9A.
20. ———— (1908). Report on the landslide at Notre-Dame de la Salette, Lièvre River. Quebec. Canada, Dept. of Mines, Geological Survey Branch.

21. FORMAN, S. A. and BRYDON, J. E. (1960). Clay mineralogy of Canadian soils. Presented to the Royal Society of Canada, Symposium on Soils in Canada, Kingston.

22. GADD, N. R. (1956). Geological aspects of Eastern Canadian flow slides. Proceedings, Tenth Canadian Soil Mechanics Conference, N.R.C., Associate Committee on Soil and Snow Mechanics, Technical Memorandum no. 46, 2–8.

23. —— (1960). Surficial geology of the Bécancour map-area, Quebec. Geol. Surv. Canada, paper 59–8.

24. HAMILTON, J. J. and CRAWFORD, C. B. (1959). Improved determination of preconsolidation pressure of a sensitive clay. Presented to the 62nd Annual Meeting, American Society for Testing Materials.

25. HODGSON, E. A. (1927). The marine clays of Eastern Canada and their relation to earthquake hazards. J. Roy. Astron. Soc. Canada, XXI, no. 7, 257–64.

26. HURTUBISE, J. E. and ROCHETTE, P. A. (1956). The Nicolet slide. Proceedings, 37th Convention of the Canadian Good Roads Association, 143–55.

27. ——, GADD, N. R., and MEYERHOF, G. G. (1957). Les éboulements de terrain dans l'Est du Canada (Landslides in Eastern Canada). Proceedings, Fourth International Conference on Soil Mechanics and Foundation Engineering, London, vol. 2, 325–9.

28. KARROW, P. F. (1960). The Champlain Sea and its sediments. Presented to the Royal Society of Canada, Symposium on Soils in Canada, Kingston.

29. KENNEY, T. C. (1959). Discussion to "The influence of rate of strain on effective stresses in sensitive clay." Presented to the 62nd Annual Meeting, American Society for Testing Materials.

30. LaFLAMME, Mgr. (1894). L'Eboulis de St. Alban. Trans. Roy. Soc. Canada, XII, Sec. IV, 63–70.

31. LEGGET, R. F. (1954). Correspondence on Paper No. 5959, Proceedings, Institution of Civil Engineers, vol. 3, no. 1, 613–6.

32. LOGAN, SIR WILLIAM (1838–42). Proc. Geol. Soc. London, 3: 767–9.

33. —— (1863). Stratified clays and sands of eastern Canada. Geol. Surv. Canada, Progress Rept. 1863, ch. 22, 915–30.

34. MEYERHOF, G. G. (1954). Field investigation of the earth-flow at Rimouski, Quebec. Proceedings, Seventh Canadian Soil Mechanics Conference, N.R.C., Associate Committee on Soil and Snow Mechanics, Tech. Mem. no. 33, 40–5.

35. MONAGHAN, B. M. (1954). The location and construction of the Quebec North Shore and Labrador Railway. Eng. J., 37, no. 7: 820–8.

36. PECK, R. B., IRELAND, H. O., and FRY, T. S. (1951). Studies of soil characteristics, the earth flows of St. Thuribe, Quebec. Soil Mechanics, Series no. 1, University of Illinois.

37. PRYER, R. W. J. and WOODS, K. B. (1959). Investigations of banded sediments along St. Lawrence North Shore in Quebec. ASTM Special Technical Publication no. 239, 55–70.

38. ROSENQVIST, I. TH. (1959). Physico-chemical properties of soils: soil-water systems. J. Soil Mechanics and Foundations Division, Proc., Amer. Soc. Civ. Eng., vol. 85, No. SM 2, pt. 1, 31–35.

39. TERASMAE, J. (1959). Notes on the Champlain Sea episode in the St. Lawrence lowlands, Quebec. Science 130: 334–6.

40. WARKENTIN, B. P. and BOZOZUK, M. (1960). Shrinking and swelling properties of two Canadian clays. Preprint of submission to the Fifth International Conference on Soil Mechanics and Foundation Engineering, Paris, 1961.

41. WILSON, M. E. (1924). Arnprior-Quyon and Maniwaki areas Ontario and Quebec. Geol. Surv., Can., Mem. 136, 9–10.

42. WOODS, K. B., PRYER, R. W. J. and EDEN, W. J. (1959). Soil engineering problems on the Quebec North Shore and Labrador Railway. Bull. Amer. Rly. Eng. Assoc. 60: 669–88.

# ENGINEERING SIGNIFICANCE OF SOILS
# IN CANADA

## R. F. Legget and R. M. Hardy, FF.R.S.C.

THE PHYSICAL DEVELOPMENT of a relatively new country such as Canada is to a very large degree dependent upon the work of the civil engineer. On the heels of the pioneer work of the first settlers must come the construction of roads, bridges and eventually railways; the provision of wharves and other aids to navigation; the installation of municipal services, water supplies and sewage disposal arrangements, as soon as settlements are large enough to require these; the building of power houses, some actuated by falling water, others by the combustion of fuel, and the associated transmission lines; and in more recent days, and particularly in Canada, the construction of airports and necessary aids to aerial navigation.

Engineering works, of which this is but a summary listing, are carried out in all parts of the country, chiefly along its southern border, but today throughout the North as well, even in the islands of the high Arctic. Disturbance of the natural ground surface is almost inevitable in all such work, and in all but exceptional cases this means disturbance of soil. The engineer always hopes to have solid bedrock available for the driving of his tunnels, and the founding of his major structures, but in almost all his work he cannot choose his site, having to accept the natural conditions at sites selected on grounds of convenience and economy. Even when bedrock is available for use, it is almost invariably covered with some "overburden."

The handling of soil in the carrying out of engineering works has therefore been a continuing feature of importance in the practice of construction. "Dirt moving" has become a term of common usage. For all too long a time, soil was regarded by engineers as a necessary nuisance, not worthy of analysis or study in the same way as other materials of construction. Fortunately, from the very earliest years of modern engineering practice, there were always exceptional individuals who did not think in this way. In France, Coulomb (5) and Collin (4) were amongst the pioneers more than a century ago. At a later date, Canada had her keen students of the engineering properties of soil, Samuel Fortier (14) being but one member of this distinguished company.

It was not, however, until the second and third decades of this century that such individual soil investigations achieved any corporate recognition. Pioneer work was done in Sweden in the investigation of serious slides in sensitive marine clays (3), but it was the publication in 1925, in the United

218

States, of a series of papers on laboratory soil studies by Dr. Karl Terzaghi (24) that proved to be the catalyst for which the profession of civil engineering had been waiting. Dr. Terzaghi has fortunately been spared to see his pioneer efforts lead to the steady development of a new scientific discipline, known generally if unfortunately as *Soil Mechanics* (a rough translation of "Erdbaumechanik").

In all parts of the world active work, both in the field and in the laboratory, now distinguishes the use of soil in engineering work. An international society brings regularly together students of soil mechanics from many nations.* As the proper engineering study of soils has developed, so has it broadened its base, so that today a new term is coming into use, embracing soil mechanics and much more, the term *Geotechnique*, one singularly appropriate for Canadian use.† That geotechnical studies embrace geology goes without saying; so also do truly effective investigations in the more restricted field of soil mechanics. In fact, laboratory soil studies carried out without the benefit of knowledge of the geological history of the samples being used may prove to be abortive; while a knowledge of local geology, and in particular of local Pleistocene geology, is an essential prerequisite to any soil exploration programme that is to be effective.

### Soils of Canada

Geology and civil engineering are thus inextricably associated in relation to all use of soil on construction, as they are correspondingly linked in relation to all engineering use of bedrock. A mere glance at the Glacial Map of Canada (10) will illustrate the significant aspects of this correlation. The almost complete absence of unglaciated areas with associated residual soils, that can be very troublesome on engineering works, is the first notable feature. The widespread existence of glacial till as the soil mantle is an equally fortunate feature, as are also the common occurrence of other glacial soil features, moraines, eskers, kames etc., providing, as they so often do, supplies of sand and gravel.

On the other side of the balance sheet, however, is the vast extent of clays deposited in the great glacial lakes of Canada, and in the encroachment of the sea up the St. Lawrence and Ottawa valleys, and around the coasts of Hudson's Bay and the eastern Arctic. Due essentially to their geological history, these clays include some of the most troublesome of all "normal" soils. Reference need be made merely to such disasters as the failure of the Transcona Elevator in 1913 (2), the tragic filling of the Beattie Mine with "mud" in 1944 and a similar accident at the Josephine Mine in 1946 (17), to demonstrate that Canada does have soils that can have

---

*International Society of Soil Mechanics and Foundation Engineering, Proceedings of Conferences.

†The word appears to have been first used in this sense in Sweden in 1913. See *Géotechnique* X, 6, 1960.

disastrous effects upon engineering works if they are not properly understood.

Canada's unusual climate makes its own contribution to engineering soil problems. Climatic factors are at least partially responsible for the vast expanses of fossilized vegetation, muskeg, that forms a surface cover over so much of northern Canada and which, although not strictly a soil, constitutes an important part of the terrain. Associated frequently with muskeg, but far more widespread and more significant, is the phenomenon of permafrost, the perennially frozen condition of the ground. When the ground consists of saturated silt or clay, the resulting permafrost can prove to be perhaps the most difficult of all soil conditions with which the engineer has to deal in the course of his operations (20).

Against this general background, brief notes on the main soil types of Canada will be presented, illustrating the significance of their geological character in relation to specific engineering works. Even though a few typical examples are all that can be cited, they will be sufficient to illustrate the interdependence of geology and soil engineering, the works of the engineer frequently revealing geological information otherwise inaccessible and sometimes, in the case of permafrost, actually modifying natural conditions. Geology alone has so far been mentioned, but the corresponding interrelation of soil engineering and pedological studies is equally close, in the case of construction work involving surface soils, as is clearly shown in the paper by Rutka in this Symposium (23).

## Glacial and Marine Clays

Crawford, in his companion paper (6), describes the results of extensive engineering tests upon the Leda Clay of the Ottawa and St. Lawrence Valleys, typical of marine clays that are described as "sensitive." This peculiar but extremely important physical property of these soils, found also with similar marine clays in Scandinavia, is directly related to the process of deposition of the soil particles that constitute these clays. That sedimentation took place in salt water is confirmed by the residual salt content in the pore water of these clays today. Variability of this salt content, through leaching, affects some of the physical properties to marked degree.

Clays formed by deposition in glacial lakes possess similar physical properties, except that they are not normally affected by residual salts in the pore water. Rominger and Rutledge (21) have reported an extensive investigation of the engineering clays of Lake Agassiz and have correlated their results with the geology of this unusually large fresh-water lake. (Their studies were conducted upon soil samples obtained in North Dakota but since the area of Lake Agassiz south of the border is a very small proportion of the total, this pioneer investigation may be considered to be almost a Canadian study.) By carrying out a variety of soil mechanics tests upon almost 450 samples from three locations, the authors were able to correlate lithologic units not easily defined by ordinary geological methods and to locate accurately the depth of an old drying surface, within the clay, that indicated a geological interval not otherwise apparent. This correlation of the results of soil mechanics tests conducted in the laboratory with geological interpretations has placed a significant new tool in the hands of the geologist, as is shown by a number of more recent studies including those of Dreimanis (7).

The unusually high natural moisture content of these clays has been shown to be due to the internal arrangement of the soil particles, the inter-molecular attraction at the time of deposition resulting in a loose, brittle structure. This structure is sensitive to disturbance and strains as low as three to five per cent may result in destruction of the internal bonds. If this occurs the soil loses its shear strength, and soils which have previously been quite "solid" in appearance and behaviour, suddenly become quite fluid (22). Crawford has mentioned in his paper the widespread landslides in the Leda Clay attributable to this cause. Similar soil instability is found with the fresh-water glacial clays. Dramatic evidence of this was provided by the slides which filled first the "glory hole" and eventually the underground workings of the Beattie gold mine near Duparquet, Quebec, in 1944 and again, and finally, in 1946. Recent studies suggest that the process of excavation being followed in excavating in the "glory hole," with a large drag-line, eventually resulted in slopes which exceeded the critical safe limit for the local glacial clay (a deposit in Lake Warren?); slope failures resulted which triggered the immense flow slides which had such tragic results (25).

A similar disaster overtook the Josephine iron ore mine, north of Sault Ste. Marie, in 1946 (9). Here the underground workings were located beneath Parks Lake which was drained of water, but not of the soil deposits in its bed, before underground mining commenced. Stoping that could not be controlled eventually broke up to the lake bottom: this disturbed the unstable glacial clay that formed the lake bed which thereupon started to flow into the workings, completely filling them in the course of three hours. The mine has not been operated since the disaster.

Innumerable minor troubles with these potentially unstable glacial lake clays illustrate the same phenomena that occurred in these major disasters, all dependent upon the geological origin of the soils in question. The typically high natural moisture content of these soils naturally affects their strength and compressibility, and so their load-carrying capacity and settlement characteristics. Today this is understood; laboratory tests upon soil samples can determine safe bearing capacities and compressibility properties in advance of design. Before these techniques were available, however, experience had to be the guide, and the solid superficial appearance of these clays led to serious differential settlements as well as to many foundation failures of which the most notable was the tilting, to an angle of 26°, of the large concrete grain elevator at Transcona, Manitoba, in 1913 when first filled with grain (2). Recent studies have demonstrated the cause of the failure, which was clearly related to the physical property of the local clay, here a deposit in Lake Agassiz.

## Glacial Till

Glacial till is usually a very compact material, its composition as a mixture ranging from boulders to clay-sized particles often giving unusually high densities. It is frequently encountered in so overconsolidated a condition that it has an almost rock-like character, until it comes into contact with water. This can be extremely serious when such material is encountered in excavation work, as recently occurred on the construction of the St. Lawrence international power and seaway project. Knowledge of the glacial history of an area can here prove to be a very useful guide, the relative age of glacial till usually indicating whether it has been consolidated by later ice pressure or not (27).

The familiar character of glacial till makes the fact more remarkable that tills have been encountered in the Ungava-Labrador area that are the reverse of overconsolidated, their solid character disappearing immediately water comes in contact with them. They are found to have an open structure, possibly due to high ice content when deposited and subsequent thawing; water can therefore quickly fill the open voids, where it is held due to the well-graded mixture of particles. As a result the till, when wet, very quickly becomes "soup," with difficulties for contractors that can readily be imagined (26).

## Some Western Soils

In Western Canada, all the foregoing soil problems are to be found as well as some that are peculiar to the west. A particularly troublesome soil type occurring widely in western and northwestern Canada is shale. Shales of Cretaceous age occur in a strip several hundred miles wide to the east of the Rocky Mountains, and may be interbedded with coal seams, silt, sand, siltstones and sandstones. These deposits are characterized by a wide variety of cementing media. Many of the shale types are compressed clays in which the major diagenetic process in their formation has been overburden pressure from the weight of overlying glaciers. They now exist at much reduced overburden pressures from the maximum during their geological history, and are therefore in a state of "rebound," in soil mechanics parlance, in which they are tending to revert to clays. The process is accelerated by the availability of subsoil moisture and weathering action, including freezing and thawing. In the field of soil mechanics such shales have been designated as "clay shales," and are also known as "preconsolidated" or "overconsolidated" soils (12).

These shales are rocks from the geological point of view but are on the border-line between rock and soil. They exhibit a very wide variation in properties of significance in engineering practice. Rock excavation techniques may be necessary for their excavation, but what appears to be hard sound rock immediately following blasting, reverts to a loose mass of clay within a few hours exposure to weathering. The rebound process under reduced overburden pressures has resulted in the formation of fractures, fissures and slickensides in the shale strata at shallow depths, particularly adjacent to the major river valleys. The parent material consists of clay-size particles varying widely in plasticity characteristics. Some deposits contain a high percentage of montmorillonite, occasionally pure bentonite, and these shales are capable of exerting high swelling pressures in the process of reverting to highly plastic clays under reduced overburden pressures.

These characteristics result in unusual problems in a wide variety of engineering practice in the areas of occurrence of such soils. In highway work, in particular, conventional methods of roadbed construction may precipitate tremendous, deep-seated slides, in which the change in stability conditions due to the construction appear to be very minor in relation to the huge forces involved in the ultimate soil movement. The recent collapse of the Peace River bridge at Taylor, B.C., is a spectacular example of the engineering problems arising with these soils. The collapse occurred some fifteen years after the bridge was built and was caused by a large slide which developed in clay shale on the north bank of the river. The slide was 900 feet long and extended to a depth of about 130 feet below the top of the shale bed. The weight of the bridge foundations in the slide area and the loads transferred to them from the bridge were insignificant compared to the forces involved in the moving mass of soil within the slide area; the bridge foundations merely rode along with the moving soil mass. The cause of the slide was a major unloading of the area adjacent to the north end of the bridge incidental to the construction of the approach road to the bridge and major alterations to the surface drainage pattern adjacent to the north end of the bridge. The precipitation during the six months before the collapse substantially exceeded the precipitation during any comparable period in the history of the bridge.

Anomalous situations can arise in building construction in the areas of these soils. In engineering practice one usually expects buildings to settle if foundation conditions are poor. With clay shales, however, the movements may be heaving rather than settlement. A second anomaly arises in connection with the concept of safe bearing pressure for the soil. Conventionally, structural designers consider that the safety of a foundation is increased by reducing the bearing pressure on the soil. For clay shales which exhibit high swelling characteristics, however, reduction in the bearing pressure may result in a less safe foundation. It is frequently advisable to specify that the bearing pressure shall not be less than a lower limit based on the swelling properties of the soil.

A large portion of Saskatchewan is underlain by the Bearpaw shale which comes within the general definition of clay shales in its characteristics. Extensive detailed investigations have been made of the engineering properties of this shale by the staff of the Prairie Farm Rehabilitation Administration of the Federal Department of Agriculture, incidental to the design of the South Saskatchewan irrigation and power project. Their work has suggested that some of the slumping along the valley of the South Saskatchewan River is due to the slow expansion of the shale both horizontally and vertically, and that these movements in recent geological time have amounted to hundreds of feet (19). It has also been shown that the rebounding shale is capable of exerting substantial swelling pressures. This characteristic is of considerable economic importance in the design of tunnels; extraordinarily high pressures build up on the linings, and the horizontal pressure has been found to be considerably greater than the vertical pressure.

The characteristics of these clay shales is a subject of current research. With the present state of engineering knowledge concerning them, it is for practical purposes impossible to predict accurately by subsurface exploration where instability will definitely develop in side hill construction. Further, the presently accepted methods for analysing the stability of slopes do not appear to be adequate to predict accurately the safety of slopes in these materials.

In the Kamloops area of British Columbia troubles have been experienced with the foundations of bulidings and irrigation structures and the performance of roads built on the local loess, a type of soil not generally realized as occurring in Canada (11). The open structure of loessal soils renders them always susceptible to the effect of water, saturation with water being a well accepted method of site preparation. In the Kamloops area, a further complication arises from the fact that the loess contains from 5 to 6 per cent of calcium carbonate, giving to the soil some artificial cohesion. Percolating water will dissolve the calcium carbonate, thus destroying the artificial cohesion and the soil will thus consolidate under its own weight together with any superimposed load that it may be supporting, with possibly disastrous results. A small colloidal clay content in such loessal soils may also impart an artificial cohesion which is destroyed by contact with water.

### Northern Soil Problems

In the north, few new soil features of significance in engineering are encountered, although the relative distribution of southern features changes markedly. Solid rock exposed on the ground surface, for example, with its elimination of all soil problems, although infrequent in southern Canada, is common in the north. Similar also, and indeed often associated with shallow rock cover, is the distribution of muskeg. In the south a minor nuisance in engineering work, except in highway construction, its distribution being so limited and spasmodic that it can usually be avoided in site selection, muskeg in the north is so extensive and can be so troublesome that its existence may make the difference between the success or failure of some engineering enterprises. When it is realised that Canadian oil companies have had to develop over 100,000 miles of trails through the "bush" of northern Alberta and British Columbia, that much of this vast mileage has been over muskeg, and that over these trails unusually heavy drilling and construction equipment has had to be moved, then it will be appreciated that "muskeg" is a very real "soil problem" to the engineer (13). It was to assist with the very practical problem of trafficability over muskeg that Dr. N. W. Radforth started his fundamental studies of organic terrain; the appearance of a paper by Dr. Radforth in this Symposium is recognition of the significant place occupied by muskeg in Canadian engineering terrain studies.

In northwestern Canada, the residual soils that are to be found in unglaciated parts of the Yukon, such as badly weathered granite, have caused minor interference with engineering work, but to no significant extent. Much older soils were encountered in test drilling work carried out recently on the Lewes River in connection with the development of power from the Whitehorse Rapids. Deep bore holes here encountered sands and silts after penetrating 150 feet of solid basalt, a surprise to some of the engineers engaged on this work, a salutary reminder to all who have heard of the occurrence of the changing geological scene.

At the nearby Whitehorse airport, engineers utilized the natural plateau provided above the town by the 200-foot deposit of glacial silt, which is here such a notable physiographic feature, to form one of the most convenient and safe of all northern airfields. Unfortunately, as all who have visited Whitehorse know, some associated engineering operations had the effect of destroying the stability that natural vegetation had given to the steep slope which separates airport and town. The subsequent erosion of this slope has become a singularly vivid reminder of the vital importance of working with nature in the protection and promotion of vegetation in all work involving soil slope stability. Arrangements are being made for this unsightly and unsatisfactory situation to be corrected.

*Permafrost*

Permafrost is encountered in the Yukon, just as in Alaska, but in the Northwest Territories it is the almost universal condition of the ground. It must be stressed that the word "permafrost" when used literally applies to a ground condition and not to a special material. Despite the valiant efforts of a geologist (Dr. Kirk Bryan) to introduce the use of semantically accurate words for the description of frozen ground, the terminological inexactitude of "permafrost" persists. Its use is further confounded by the singular semantic sloppiness of engineers who sometimes use the word to describe perennially frozen silt, the major cause of trouble to engineers in areas of permafrost.

At a depth of about twenty feet the temperature of undisturbed ground can be shown to be reasonably constant throughout the year. Between this depth and the surface the temperature of the ground varies. The closer to the surface, the greater is the variation. It is generally seasonal but at shallow depths it is also diurnal. The steady temperature at depth is usually found to be within a few degrees of the mean air temperature for the locality throughout the year. When, therefore, local annual mean air temperature is appreciably below 32° F, the ground will be perennially frozen, thawing occurring only very close to the surface during later weeks of the short northern summers. At depths below twenty feet, ground temperature is known to increase slowly but steadily, in places at the rate of one degree for every 150 feet. It therefore follows that with perennially frozen ground, the frozen condition will persist downwards to such a depth that the normal increase of temperature with depth eventually raises the sub-freezing ground temperature to 32° F.

Permafrost in Canada may therefore be visualized as an immense wedge-shaped condition of the ground, the thin end of the wedge being a very irregular southern band of discontinuous frozen ground, the wedge thickening towards the Pole. Maximum depth so far observed in Canada is approximately 1,300 feet at Resolute Bay on Cornwallis Island. Although there are scattered isolated references to "frozen ground" in the older literature of the north, some of unusual interest, the significance of permafrost was not generally appreciated until the last two decades. Recent geological interest in geothermal problems has directed the attention of some geologists to the fascinating theoretical questions posed by permafrost and its manifestations, but it has been the work of the engineer in opening up the north, starting with the construction of the Alaska Highway, that has focused attention upon the practical problems created, in some areas, by permafrost. As is so often the case, the search for answers to these problems has led first engineers and then scientists into quite basic studies of the thermal regime of the surface of the ground, with special reference to the north (16).

This general statement on permafrost appears desirable in this context, not only because permafrost appears to have been previously mentioned only incidentally to the Royal Society of Canada, but also because, without a clear appreciation of the over-all picture of permafrost, any brief reference to the frozen soil problems of the north will not be too meaningful. When the ground consists of solid rock or well-drained and dry soils, such as sands and gravels, the frozen condition is generally of theoretical interest only. When, however, permafrost is the condition of saturated silts or clays, it becomes of major practical importance. If the frozen condition is ameliorated, material previously rock-like in character may be quickly converted into the engineering equivalent of soup, with results that can be generally imagined. Thus, when muskeg was scraped off frozen glacial silt in early northern road construction, the hard frozen silt quickly became a quagmire. Heated buildings in contact with frozen soil would start to settle, and not always uniformly, as soon as their heating systems were placed in operation. Even the effect of solar radiation on large buildings in the far north has been found to cause differential settlements, to be measured in feet, between the north and south facing walls.

These were early engineering problems; solutions to them are now available: study of them continues. What are the geological implications of permafrost? Its origin is a geological problem of prime importance. The geothermal field investigations that are now proceeding will shed light not only on this problem but on other questions relating to the earth's thermal regime in general. As excavations in frozen ground proceed, in the course of northern engineering work, it is greatly to be hoped that watch will be carefully maintained for unusual geological and other features that may be thus revealed. In recent excavations in northern Manitoba varved silts were exposed that revealed an unusual pattern of ice-lensing: this is still being studied but will be eventually recorded in some suitable medium. Similarly, occurrences of ground ice in the Aklavik area, N.W.T., and of unusual cavities, probably due to the melting of frozen boulders, found in frozen silt near Uranium City, Saskatchewan, are permafrost features of geological significance that would have escaped notice had it not been for the prosecution of engineering work.

*Engineering Excavations*

The contributions that may be made to geological knowledge, and particularly to the extension of Pleistocene information in Canada, through careful observation of the soil cross sections revealed by engineering excavation work, is the natural and welcome reciprocal of the contributions that geological information can make to the satisfactory conduct of soil engineering. Notable in Canadian experience of this kind was the use made of the remarkable exposures of varved glacial clays in the bed of Steep Rock Lake, after draining.

This lake in western Ontario had an area of about ten square miles. It formed part of the course of the Seine River but the river was diverted by a major civil engineering operation so that two ends of the lake could be dammed and the entire lake pumped out. This operation was successfully completed. Subsequent excavation of the lake bed deposits revealed what is probably the largest exposure of varved glacial clays ever seen. These were studied by Dr. Antevs (1) as purely geological phenomena and also in connection with the engineering development of the two iron ore mining operations that are now in progress in the old lake bed (8). Geological knowledge of varved material was thus advanced almost as a by-product of engineering operations (15).

On a different scale and with entirely different geological conditions, the excavation for Canada's first subway, the Yonge Street line of the Toronto Transit Commission, was similarly used for geological purposes incidental to the major engineering undertaking represented by the construction of this subway line. A complete soil profile along the subway was obtained and a full suite of soil samples was deposited for future use by geologists in the Royal Ontario Museum. Geological studies of the soils thus revealed have added further information to the stock of knowledge regarding the justly famous Toronto interglacial beds (18).

## Conclusion

This general review of some of the soil problems associated with civil engineering work throughout Canada, and of some contributions to the geological appreciation of Canadian soils contributed indirectly by this enginering work, will have made clear the complete dependence of much civil engineering upon adequate knowledge of soil conditions. Geologists and pedologists have made notable contributions to the success of engineering undertakings. It is satisfactory to record that some return has been made by the geological information revealed by engineering excavation. With the growing appreciation of the importance of geology in all engineering work it may be expected that this trend will continue to the mutual benefit of engineering and geology.

REFERENCES

1. ANTEVS, ERNST (1951). Glacial clays in Steep Rock Lake, Ontario. Bull. Geol. Soc. Amer., *62*: 1223–62.
2. BARACOS, A. (1957). The foundation failure of the Transcona grain elevator. Eng. J., *40*: 973.
3. BJERRUM, L. and FLODIN, N. (1960). The development of soil mechanics in Sweden, 1900–1925. Norwegian Geotechnical Institute, Oslo, publication *36*: 1–18.
4. COLLIN, ALEXANDRE (1846). Landslides in clays. Translated by W. R. Schriever, University of Toronto Press, 1956.
5. COULOMB, C. A. (1773). Essai sur une application des règles de maximis et minimis à quelques problèmes de statique, relatifs à l'architecture. Paris, Mémoires de l'Académie des Sciences (Savants Etrang.) *7*: 343–82.
6. CRAWFORD, C. B. Engineering studies of Leda clay. Presented to the Royal Society of Canada, Symposium on Soils, Kingston, 1960. See above, p. 200.
7. DREIMANIS, A. Tills of Southern Ontario. Presented to the Royal Society of Canada, Symposium on Soils, Kingston, 1960. See above, p. 80.
8. EDEN, W. J. (1955). A laboratory study of varved clays from Steep Rock Lake, Ontario. Amer. J. Sci., *253*: 659–74.
9. GALLIE, ALAN E. (1947). Mining methods and costs at the Josephine mine. Bull. Can. Inst. Min. Metall., *427*: 589–683.
10. Glacial map of Canada (1958). Geological Association of Canada, Toronto.
11. HARDY, R. M. (1950). Construction problems in silty soils. Eng. J., *33*: 775–9.
12. ——— (1957). Engineering problems involving preconsolidated clay shales. Trans. Eng. Inst. Can., *1*: 1–10.
13. KEELING, L. (1958). Some aspects of the terrain problems in Northwestern Canada. N.R.C., Associate Committee on Soil and Snow Mechanics, 4th Muskeg Research Conference, Proceedings, technical memorandum 54: 11.
14. LEGGET, R. F. (1949). Samuel Fortier, pioneer in soil mechanics. Eng. J. *32*: 68.
15. ——— (1958). Soil engineering at Steep Rock iron mines, Ontario, Canada. Proc. Inst. Civ. Eng., *11*: 169–88.
16. ——— , DICKENS, H. B., BROWN, R. J. E. Permafrost investigations in Canada. First International Symposium on Arctic Geology (in press).

17. —— and EDEN, W. J. (1960). Soil problems in mining on the Precambrian Shield. Eng. J. *43*, 81–7.
18. —— and SCHRIEVER, W. R. (1960). Site investigations for Canada's first underground railway (the Toronto subway). Civil Engineering and Public Works Review, *55*: 73.
19. PETERSON, R. (1958). Rebound in the Bearpaw shale, Western Canada. Bull. Geol. Soc. Amer., *69*: 1113–24.
20. PIHLAINEN, J. A. and JOHNSTON, G. H. (1954). Permafrost investigations at Aklavik (drilling and sampling), 1953. N.R.C., Division of Building Research, publication NRC 3393.
21. ROMINGER, J. F. and RUTLEDGE, P. C. (1952). Use of soil mechanics data in correlation and interpretation of Lake Agassiz sediments. J. Geol., *60*: 2.
22. ROSENQVIST, I. TH. (1959). Physico-chemical properties of soil: soil water systems. Proc. Amer. Soc. Civ. Eng., *58*, S.M. 2: 31–54.
23. RUTKA, A. Correlation of engineering and pedological soil classification in Ontario. Presented to Royal Society of Canada, Symposium on Soils, Kingston, 1960. See above, p. 183.
24. TERZAGHI, KARL (1925). Principles of Soil Mechanics. Engineering News-Record, *95*.
25. TUTTLE, JAY (1939). The spiral stoping system as applied at the Beattie mine. Trans. Can. Inst. Min. Metall., *42*: 95–122.
26. WOODS, K. B., PRYER, R. W. J., and EDEN, W. J. (1959). Soil engineering problems on the Quebec North Shore and Labrador Railway. Amer. Rly. Eng. Assoc. Bull., *60*: 669–88.
27. WRIGHT, G. F. (1920). Ice age in North America and its bearing on the antiquity of man. 6th ed. Oberlin Publishing Company, Ohio.

# INDEX

## II. Index of Place Names

## III. Subject Index